Precaution matters, especially when we may be on the brink of passing tipping points fit to cause catastrophic and irreversible climate change. This book does an admirable job of making the right distinctions in the right places, so as to enable a better understanding of what precaution means in the mess we are in. The distinctive view of precaution defended in the book – in particular, the Catastrophic Precautionary Principle and the Catastrophic Precautionary Decision-Making Framework – moves climate politics forward in novel and much needed ways.

Catriona McKinnon, Professor of Political Theory,
University of Reading

How can we get a sane grip on the real possibility that extreme climate change will unleash catastrophe? This highly original, widely knowledgeable, and deeply powerful argument shows through a balanced but revealing analysis of the three core approaches of Nordhaus, Stern, and Wagner & Weitzman that the social cost of carbon is being systematically underestimated because of blind-spots in the fundamental assumptions of economics that inevitably mask uncertain dangers of catastrophe that public policy neglects at peril to many generations. A wise and exceptionally important book accessible to non-specialists.

Henry Shue, Senior Research Fellow, Centre for International Studies,
University of Oxford, and author of Climate Justice

A Climate of Risk

We are living in a climate of risk. Our way of life imposes risks on ourselves and others. We are causing climatic changes that have the potential to change radically the conditions under which both we – the present generation – and future generations will live. While we are now quite certain that climate change is happening, we are unsure of exactly what will happen and when, given different emissions and policy scenarios. We are therefore in a position where we must decide what to do about the risks climate change threatens in the face of a range of uncertainties.

In this book, Lauren Hartzell-Nichols provides guidance in the face of this uncertainty by offering an in-depth discussion of how and why we ought to take a precautionary approach to climate policy, namely by appeal to a Catastrophic Precautionary Principle and Catastrophic Precautionary Decision-Making Framework. By examining the way in which climate change is harmful, Hartzell-Nichols shows how precaution does have a meaningful role to play in moving climate policy forward if we reconsider what precaution is about before too quickly appealing to precaution as a reason or justification for action.

A Climate of Risk takes a philosophically grounded, interdisciplinary approach that will appeal to a broad scholarly and policy-oriented audience. Hartzell-Nichols's reinterpretation of the precautionary principle enables precaution to be more effectively leveraged as a driver of action on climate change.

Lauren Hartzell-Nichols is an affiliate assistant professor of philosophy at the University of Washington. She has published widely on many topics in climate ethics, including precaution, adaptation, and geoengineering, both on her own and with diverse, interdisciplinary teams. Her work addresses the ethical challenges climate change poses. In particular, she addresses the complexity of ethical decision making in the face of significant, intergenerational risks.

Environmental Politics / Routledge Research in Environmental Politics

Edited by Steve Vanderheiden
University of Colorado at Boulder

Over recent years environmental politics has moved from a peripheral interest to a central concern within the discipline of politics. This series aims to reinforce this trend through the publication of books that investigate the nature of contemporary environmental politics and show the centrality of environmental politics to the study of politics per se. The series understands politics in a broad sense and books will focus on mainstream issues such as the policy process and new social movements as well as emerging areas such as cultural politics and political economy. Books in the series will analyse contemporary political practices with regards to the environment and/or explore possible future directions for the 'greening' of contemporary politics. The series will be of interest not only to academics and students working in the environmental field, but will also demand to be read within the broader discipline.

The series consists of two strands:

Environmental Politics addresses the needs of students and teachers, and the titles will be published in paperback and hardback. Titles include:

Global Warming and Global Politics
Matthew Paterson

Politics and the Environment
James Connelly & Graham Smith

International Relations Theory and Ecological Thought
Towards Synthesis
Eric Laferrière & Peter Stoett

Planning Sustainability
Edited by Michael Kenny & James Meadowcroft

Deliberative Democracy and the Environment
Graham Smith

EU Enlargement and the Environment
Institutional change and environmental policy in Central and Eastern Europe
Edited by JoAnn Carmin and Stacy D. VanDeveer

The Crisis of Global Environmental Governance
Towards a new political economy of sustainability
Edited by Jacob Park, Ken Conca and Matthias Finger

Routledge Research in Environmental Politics presents innovative new research intended for high-level specialist readership. These titles are published in hardback only and include:

A Climate of Risk

Precautionary Principles, Catastrophes, and Climate Change

Lauren Hartzell-Nichols

Routledge
Taylor & Francis Group

LONDON AND NEW YORK

First published 2017 by Routledge

2 Park Square, Milton Park, Abingdon, Oxfordshire OX14 4RN
52 Vanderbilt Avenue, New York, NY 10017

Routledge is an imprint of the Taylor & Francis Group, an informa business

First issued in paperback 2019

Library of Congress Cataloging in Publication Data
Names: Hartzell-Nichols, Lauren, author.
Title: A climate of risk : precautionary principles, catastrophes, and
    climate change / by Lauren Hartzell-Nichols.
Description: New York : Routledge, [2017] | Series: Environmental
    politics/Routledge research in environmental politics ; 26 | Includes
    bibliographical references and index.
Identifiers: LCCN 2016045683 | ISBN 9781138233577 (hbk)
Subjects: LCSH: Climatic changes--Prevention. | Global environmental
    change. | Precautionary principle.
Classification: LCC QC903 .H38 2017 | DDC 363.738/74--dc23
LC record available at https://lccn.loc.gov/2016045683

ISBN: 978-1-138-23357-7 (hbk)
ISBN: 978-0-367-37176-0 (pbk)

Typeset in Times New Roman
by Taylor & Francis Books

Contents

Figures

Acknowledgements

The history of this book is the history of my intellectual development. Attempting to pinpoint its origin or when I started working on it feels as nebulous as trying to identify the moment climate change became anthropogenic climate change.

Apparently I never stopped asking "why?" in the incessant way my three-year-old does now. I do vividly remember the experience – a backpacking trip the summer before my junior year of high school – that led me to suddenly feel the enormity of time and nature in a way that pushed me to think in terms of generations rather than decades, years, or minutes. I also remember having to sheepishly admit to my mom that she was right, I did love my first philosophy course, which I only took my first semester of college to get it out of the way and prove her wrong. My first philosophical obsession was then quite naturally the non-identity problem and our ethical relationship with future people – the appendix to this book contains glimmers of my undergraduate honors thesis, my first attempt to grapple with this subject in a meaningful way.

I stumbled into the topic of precaution the summer I interned for my undergraduate advisor, Derek Turner, thanks to a generous program at Connecticut College. Derek treated me as a true collaborator as we wrote the paper that would seed my dissertation and then this book. (Derek has forever since been one of my most trusted mentors and has helped me at every stage of the development of this book.) Only I never intended for it to be that way: I tried – and failed – to propose several other dissertation topics before I found myself returning to precaution. Despite being at a school (Stanford) that did not have any resident experts on environmental or climate ethics, I managed to put together a committee of brilliant scholars who collaboratively supported my (then) rather non-conventional project. I hence owe huge debts of gratitude to Joshua Cohen, Tamar Schapiro, (the late) Stephen Schneider, and above all Debra Satz (my dissertation chair).

Stephen Gardiner, Michael Blake, Allison Wylie, and Kristin Shrader-Frechette were all instrumental in guiding me as I set out determined to transform my dissertation into a book. They wisely suggested (or supported) that I take my time to work out my ideas and develop as an intellectual by publishing in academic journals before rushing to publish a book. (Of course, I do not think any of them expected it would take me another seven years to get to this point.) In the end my dissertation served more as a launch pad to this book project than its core, as my ideas and the body of scholarship on climate change have advanced over the years. Stephen especially, but also Michael, have been extremely generous with their time, mentoring me and reading drafts of papers, chapters, and my entire manuscript. Henry Shue and Daniel Steel also graciously agreed to read an early draft of my manuscript, and I am grateful for their helpful comments. Paul Kelleher pushed me to rethink and revamp my analysis of the economics of climate change in a way that, I think, elevated my engagement in important ways.

So many others also helped me in large and small ways, especially my collaborators on related projects – Kirsten Oleson, Michael Mastrandrea, Daniela Cusack, John Axsen, Rachael Shwom, Sam White, Katherine Mackey, Gwen Ottinger, and Timothy Hargrave – and the audiences and commentators at the many venues where I have presented the work on which this book is built. Along with the anonymous reviewers of this book, all those mentioned helped me bring to fruition what you find here. Since this book is the realization of everything I have learned and worked on to date, virtually everyone with whom I have intellectually engaged has helped this book become what it is.

I am a better philosopher and this is a better book because of all of the amazingly generous support and mentorship I have received over the years, but I could not (or, as importantly, would not) have reached this point without the unrelenting support of my friends and family. My mom and sisters (Maddie McAlister, Tia Radant, and Zoe Wisnoski) are and have been the foundation for everything I have done and built (my father, Tom Hartzell, having passed away when I was just nine). My husband, Isaac, has witnessed and supported the final leg of this project, from dissertation to long-awaited book, something that took great patience and dedication. Finally, bringing our son, Harvey, into this world changed my perspective in a way that gave new meaning to my understanding of and sense of urgency about the climate of risk in which we find ourselves. He is now the most important part of my world and grounds my hopes and fears about our collective future.

Credits

Sections of this book are in part derived from the following publications:

Lauren Hartzell-Nichols. 2014. "Adaptation as Precaution." *Environmental Values* 23(2): 149–64, doi:10.3197/096327114X13894344179121.

Lauren Hartzell-Nichols. 2014. "The Price of Precaution and the Ethics of Risk," Ethics, Policy & Environment 17(1) (2014): 116–18, available online: http://dx.doi.org/10.1080/21550085.2014.885183

Lauren Hartzell-Nichols. 2013. "How is Climate Change Harmful?" *Ethics & the Environment* 17(2): 97–110, doi:10.1353/een.2012.0017.

Lauren Hartzell-Nichols. 2013. "From 'the' Precautionary Principle to Precautionary Principles." *Ethics, Policy & Environment* 16(3): 308–20, doi:10.1080/21550085.2013.844569. First published in *Ethics, Policy & Environment* on 15 October 2013, available online: http://dx.doi.org/10.1080/21550085.2013.844569

Lauren Hartzell-Nichols. 2012. "Precaution and Solar Radiation Management." *Ethics, Policy & Environment* 15(2): 158–71, doi:10.1080/21550085.2012.685561. First published in *Ethics, Policy & Environment* on 25 July 2012, available online: http://dx.doi.org/10.1080/21550085.2012.685561

Lauren Hartzell-Nichols. 2012. "Intergenerational Risks." In *Handbook of Risk Theory: Epistemology, Decision Theory, Ethics, and Social Implications of Risk*, ed. Sabine Roeser et al., 931–60. New York: Springer.

Lauren Hartzell-Nichols. 2011. "Responsibility for Meeting the Costs of Adaptation." *Wiley Interdisciplinary Reviews: Climate Change* 2(5) (6 September): 687–700, doi:10.1002/wcc.132.

Abbreviations

a.k.a.	also known as
AMOC	Atlantic meridional overturning circulation
AR	Assessment Report
CO_2	carbon dioxide
DICE	Dynamic Integrated model of Climate and the Economy
GDP	gross domestic product
GHGs	greenhouse gases
GtC	gigatons of carbon
IPCC	Intergovernmental Panel on Climate Change
PAGE	Policy Analysis of the Greenhouse Effect model
ppm	parts per million
RCP	Representative Concentration Pathway
RFC	reasons for concern
UNFCCC	United Nations Framework Convention on Climate Change

Introduction

We are living in a climate of risk. Our way of life imposes risks on ourselves and others. In particular, we are causing climatic changes that have the potential to radically change the conditions under which both we – the present generation – and future generations will live. Each passing year brings more dire news about our increasingly warming climate. The most unsettling reports to date suggest that irreversible changes are already taking place that threaten to have catastrophic consequences. Yet the climatic changes that we are witnessing are just the very tip of the iceberg. Many of the greenhouse gases (GHGs) that have been emitted to this point will continue to affect the climate for thousands or even tens of thousands of years.[1] This means that our climate-affecting activities have the potential to harmfully impact very distant future generations. So while it may be tempting to think that we have moved beyond having to think about climate change in terms of risks and uncertainties, this is far from the case. We are now quite certain *that* climate change is happening, but uncertainty abounds in our understanding of exactly *what* will happen, given different emissions and policy scenarios. We know climate change will be bad, but we do not know exactly *how* bad it will be *when* and *for whom*. We are therefore in a position where we must decide what to do about the risks climate change threatens in the face of a range of uncertainties.

For example, melting ice sheets will increasingly become a major driver of sea level rise, but significant uncertainty remains about the rate and timing of melting for different ice sheets. Until recently we were genuinely uncertain about the likely temperature threshold at which the West Antarctic Ice Sheet would begin melting. So it came as a surprise to many when evidence revealed that the complete loss of this ice sheet is quite possibly inevitable over the next millennium.[2] This melting will contribute about an additional 10 feet to sea level rise.[3] The timing and rate of melting and the actions we take to adapt to our changing coastlines will both affect how harmful this will be. Very gradual melting will likely be far less harmful than very rapid melting because it will take time to successfully adapt to changing coastlines (e.g., think of the effect of 5 feet of sea level rise on New York City over 15 years as opposed to 150 years). Whether we take precautionary measures in advance of sea level rise (e.g., build sea walls) or wait to address the implications of rising waters when it happens (e.g., have to flee inundated cities) will affect the lives of millions or even billions of people. Hence despite our growing knowledge about climate change we must decide how to respond to the threat(s) of melting ice sheets – and all possible climate impacts – under conditions of significant uncertainty.

This book aims to provide direction in the face of this uncertainty. It explores the extent to which the precautionary principle may provide the kind of guidance we need to address the certain yet also uncertain threats of harm posed by climate change. The precautionary principle is generally understood as capturing the intuition that it is better to be safe than sorry and/or that we sometimes ought to act in advance of scientific certainty. It has long been understood as a possible guide to decision making in the face of uncertainty. Neither the content nor nature of this principle, however, has been clearly or consistently identified or applied in either environmental policy or the academic literature, and while the precautionary principle was once a major driver of climate policy and is still prevalent in diverse policy contexts, it does not currently play a significant role in the United Nations Framework Convention on Climate Change (UNFCCC). Yet intuitively climate policy is or should be all about precaution. It aims to guide us to prevent or minimize possible climate damage in the face of an incredibly complex phenomenon, our understanding of which inherently and contextually involves significant uncertainties. A major goal of this book is to show that precaution *does* have a meaningful role to play in moving climate policy forward, but in order for this to be the case we have to reconsider what precaution is and is all about before too quickly appealing to precaution as a reason or justification for action.

Chapter 1 unpacks how and why it is challenging to understand the ways in which climate change is and will be harmful. Not only is climate change a complex physical phenomenon, but also the fact that it operates over very long timescales further complicates how we should understand the ways in which it is harmful. By providing a way of interpreting what we mean when we say climate change is harmful (many of the details of which are presented in the Appendix), Chapter 1 provides a foundation for understanding why precaution is an appropriate concept to apply to climate change.

Chapter 2 then goes on to make sense of what precaution is all about. Wading through the vast literature on the precautionary principle reveals that there is no one precautionary principle or way of strongly unifying all work on precaution. Nonetheless, there are many lessons to be learned about how, when, and why precaution can be a powerful concept. For example, particular precautionary principles that address a limited context in which precaution applies may be formulated so as to avoid many of the worries about lack of clarity plaguing many so-called versions of the precautionary principle. There is much to be learned about making decisions in contexts of uncertainty from those who have viewed the precautionary principle as guiding us to a new paradigm of decision making in such contexts.

Chapter 3 defends one such precautionary principle and approach to precautionary decision making that applies to climate change, namely a Catastrophic Precautionary Principle and Catastrophic Precautionary Decision-Making Framework. While other precautionary approaches may apply to other aspects of climate policy, the driving reason why we ought to be aggressively addressing climate change is because it threatens to severely and harmfully affect many, many millions (if not billions) of people. Making explicit our reasons for taking a precautionary approach to threats of catastrophe hence has the potential to powerfully ground and motivate a precautionary approach to climate policy. We have very strong pro tanto moral reasons for taking precautionary measures against threats of catastrophe; hence we have very strong pro tanto moral reasons for taking a very aggressive precautionary approach to climate policy.

While Chapter 3 provides the foundational arguments for the Catastrophic Precautionary Principle and the Catastrophic Precautionary Decision-Making Framework, Chapter 4 defends this precautionary approach against claims that cost-benefit analysis or economics in general supplants the need for a separate precautionary principle. We need both guiding moral principles and ethically informed economic assessments of our options because, while relevant, economic considerations should not by default trump moral reasons. The Catastrophic Precautionary Principle demands that we not sacrifice moral obligation to economic interest, though economic analyses can inform our precautionary assessments of threats of catastrophe. Uncertainty, however, presents serious challenges to economic assessments of climate change. Looking at a range of such assessments in Chapter 4 reveals that some more appropriately accommodate uncertainty than others, though none is fully satisfactory from the precautionary perspective defended in this book.

Chapter 5 finally describes what is entailed in taking a precautionary approach to climate policy guided by the Catastrophic Precautionary Principle and Catastrophic Precautionary Decision-Making Framework: mitigation efforts should aim to stabilize atmospheric carbon dioxide (CO_2) and other GHGs as low as possible; adaptation policy should aim to minimize the harmfulness of eventual climate impacts in a way that simultaneously promotes other social goals such as sustainable development (where this entails a globally coordinated approach to adaptation); and geoengineering strategies should be considered as possible additional responses so long as they do not create or exacerbate threat(s) of catastrophe themselves. Other moral and political principles will have to come into play to determine exactly what strategies should be implemented and by whom to meet these goals, but committing to this precautionary approach will make the demandingness of climate policy clear. It is unacceptable from a moral standpoint that many people continue to argue that stabilizing global mean surface temperatures at 2°C (or even 1.5°C) above preindustrial levels would be enough to prevent dangerous anthropogenic influence on the global climate. Integrating the Catastrophic Precautionary Principle into our guiding climate policies, such as the UNFCCC, may provide the strong backbone needed to ground a more aggressive approach to climate policy than has been realized to date. This book is meant to help us do what we should have been doing for a long time now: implement effective precautionary climate policies in a real effort to avert climate catastrophe.

Notes

1 Ciais et al. 2013; Archer 2008.
2 Joughin et al. 2014; Rignot et al. 2014; Sumner 2014.
3 Meehl et al. 2007, 819.

References

Archer, David. 2008. *The Long Thaw: How Humans are Changing the Next 100,000 Years of Earth's Climate.* Princeton: Princeton University Press.

Ciais, P., C. Sabine, G. Bala, L. Bopp, V. Brovkin, J. Canadell, A. Chhabra, et al. 2013. "2013: Carbon and Other Biogeochemical Cycles," *Climate Change 2013: The Physical Science Basis. Contribution of Working Group I to the Fifth Assessment Report of the Intergovernmental Panel on Climate Change*, 465–570. doi:10.1017/CBO9781107415324.015.

Joughin, Ian, Benjamin E. Smith, and Brooke Medley. 2014. "Marine Ice Sheet Collapse Potentially Under Way for the Thwaites Glacier Basic." *Science* 344 (6185): 735–738.

Meehl, G.A., T.F. Stocker, W.D. Collins, P. Friedlingstein, A.T. Gaye, J.M. Gregory, A. Kitoh, et al. 2007. "2007: Global Climate Projections." In *Climate Change 2007: The Physical Science Basis. Contribution of Working Group I to the Fourth Assessment Report of the Intergovernmental Panel on Climate Change*, edited by S. Solomon, D. Qin, M. Manning, Z. Chen, M. Marquis, K.B. Averyt, M. Tignor, and H.L. Miller. Cambridge: Cambridge University Press.

Rignot, E.J., J. Mouginot, M. Morlinghem, H. Seroussi, and B. Scheuchl. 2014. "Widespread, Rapid Grounding Line Retreat of Pine Island, Thwaites, Smith, and Kohler Glaciers, West Antarctica, from 1992 to 2011." *Geophysical Research Letters* 34(2): 3502–3509.

Sumner, Thomas. 2014. "No Stopping the Collapse of West Antarctic Ice Sheet." *Science* 344(6185): 683.

1 How Climate Change is Harmful

We are worried about climate change because it will be harmful. Climate change creates a climate of risk that threatens to harmfully affect not only the present generation but potentially all future generations as well. But *how* is climate change harmful and to *whom*? This, it turns out, is a surprisingly difficult question to answer. To begin with, while it may seem obvious that people will be harmfully affected by climatic changes (e.g., by rising seas), the complex nature of climate change combined with uncertainties in our understanding of it makes it challenging to attribute specific harmful outcomes to climate change. Further, the intergenerational dimension of climate change – the time delays between causes and effects – challenges our understanding of the ways in which our climate-affecting activities harmfully affect those who will be negatively impacted by climate change.[1] At the same time, it is very intuitive to think of climate change as being harmful. While this chapter will suggest that the way in which climate change is harmful is often misunderstood, it seems worth holding on to the idea that climate change is and will be harmful to many of the individuals who are affected by it, even if we need to revise our use of the concept of harmfulness in this context.

To be able to offer an account of how we should approach ethics in a climate of risk we therefore need a way to understand and talk about the harmful implications of the risks climate change poses. I argue that when we say someone is harmfully affected by climate change we mean something between what we mean when we say someone was harmed by being struck by lightning and what we mean when we say someone was harmed by being shot by a sniper. That is, climate change is anthropogenic – caused by humans – but the causal forces at work are very complex and operate over very long time horizons, so when someone is negatively affected by climate change this is the result of a complex suite of factors, none of which can be tied back to any single climate-affecting human activity directly. Acts or policies that contribute to climate change are not harmful in the usual here-and-now way, but such actions impose threats of harmful conditions upon future people. Importantly, our standard intuitions about here-and-now harming do not apply to cases like climate change that span many generations. Of course this does not mean that there is nothing wrong with our climate-affecting activities that threaten to harmfully affect both the present and future generations, rather it means we need to carefully reflect on why we intuitively think the ways in which climate change are and will be harmful are bad and what this implies about what we ought to do to address climate change.

This chapter lays the foundation for an exploration of these bigger issues by clearly identifying the ways in which climate change is harmful. The first section provides a brief discussion of climate change and its possible effects. The second section explores the ways in which the complexity of the climate system can challenge our understanding of how climate change is harmful. The third section addresses the ways in which the intergenerational nature of climate change dominates our ability to understand the ways in which climate change is harmful and proposes a way of understanding what we mean when we say climate change is harmful.

1 Climate Change Risks, Reasons for Concern, and the Complexity of Harmfulness

Earth's climate has changed dramatically throughout its history. Up until recently these changes were the result of many naturally occurring processes. Anthropogenic climate change, however, is occurring because the actions of humans are changing the Earth's climate system. Understanding the physical processes of climate change requires understanding the relationship between atmospheric greenhouse gas (GHG) concentrations and mean global surface temperatures as well as the relationship between GHG emissions and atmospheric GHG concentrations.

The first relationship has to do with what causes temperatures to rise on a global scale. The sun provides Earth with solar energy. How much of the solar energy the Earth reflects, emits, and absorbs is based on properties of the Earth's atmosphere and surface. Changing any or all of these factors and processes affects how much of the sun's energy the Earth receives and captures, its energy budget, which in turn affects the average global temperature. There is solid scientific evidence that the Earth's energy budget has changed since the pre-industrial era. A key factor in the altered energy budget is the concentration of GHGs in the atmosphere. These gases prevent energy from escaping that would otherwise have reflected back into space. Another key factor changing the energy balance is aerosols in the atmosphere that both reflect and absorb solar radiation. The effect of these two factors on the energy balance is termed "radiative forcing." Positive forcing warms the Earth and negative forcing cools it. Anthropogenic influences are causing both positive and negative forcings on the climate (e.g., increased GHG concentrations trap more energy while higher aerosols reflect more energy). There are also numerous positive and negative feedbacks within the climate system, but the balance has shifted decidedly towards positive forcing. As the Intergovernmental Panel on Climate Change (IPCC) famously stated, "[w]arming of the climate system is unequivocal."[2] We are living in a changing climate, and we are the drivers of this change.

The second important relationship to understand is between GHG emissions and atmospheric GHG concentrations. The dominant contributor to positive forcing is increased atmospheric GHG concentrations, and these increases are predominantly the result of a long history of human activities such as fossil fuel consumption, cement manufacture, land use change, and agriculture. Current atmospheric GHG concentrations "have increased to levels unprecedented in at least the last 800,000 years," having risen by 40% since pre-industrial times.[3] These figures reflect the fact that to date, the "ocean has absorbed about 30% of the emitted anthropogenic carbon dioxide, causing ocean acidification."[4] Also significantly, GHG emissions do not immediately impact the climate but rather the accumulation of GHGs in the atmosphere causes climatic changes over time. While greenhouse gases mix in the atmosphere within weeks of being emitted, it takes decades to millennia for the full effects of increased GHGs to take hold. GHG molecules can persist in the atmosphere for long periods of time, some for thousands – or even hundreds of thousands[5] – of years. GHG emissions hence have prolonged influence on global climate.

These changes are leading and will lead to further direct and indirect human impacts. Direct impacts include deaths, injuries, disease, and other harmful effects that climate change directly causes such as from sea level rise or reduced precipitation. Indirect impacts are effects caused by stress on systems from climate change. Examples of indirect impacts include the effects of economic impacts, stresses on health care systems, and migration due to impacts on agricultural systems. Indirect impacts will probably affect far more people than direct impacts.

The concept of dangerous climate change is a helpful starting point for thinking about how and why climate change will be harmful. This concept originated in the United Nations Framework Convention on Climate Change (UNFCCC).[6] The Framework establishes that parties to the Framework Convention should avoid "dangerous anthropogenic interference" with the climate system, though they do not define this concept there. As the IPCC recognizes, defining dangerous climate change is an inherently normative task because which climate change impacts we think are dangerous depends on the ways in which they will negatively impact things we find valuable.[7] Two important concepts provided in the IPCC for understanding dangerous climate change are key risks and reasons for concern.

The idea of there being "reasons for concern" (RFC) about climate change was introduced in the Third Assessment Report (AR3) as a way to capture key risks and vulnerabilities in five categories: (1) unique and threatened systems, (2) extreme weather events, (3) distribution of impacts, (4) global aggregate impacts, and (5) large-scale singular events.[8] These reasons for concern were reaffirmed in the IPCC's Fourth (AR4) and Fifth

(AR5) Assessment Reports: "The identification of potential key vulner-abilities is intended to provide guidance to decision-makers for identifying levels and rates of climate change that may be associated with 'dangerous anthropogenic interference' (DAI) with the climate system, in the termi-nology of United Nations Framework Convention on Climate Change (UNFCCC) Article 2."[9] AR5, however, also introduced the concept of key risks to capture risks associated with high hazard and/or vulnerability of societies and systems. Key risks are defined as "potentially severe adverse consequences for humans and social-ecological systems resulting from the interaction of climate-related hazards with vulnerabilities of societies and systems exposed."[10] Looking at these key risks makes it clear that climate change threatens to be harmful in diverse ways. These are the:

- Risk of death, injury, and disruption to livelihoods, food supplies, and drinking water, in addition to loss of common-pool resources, sense of place, and identity due to sea level rise, coastal flooding and storm surges affecting high concentrations of people, economic activity, bio-diversity, and critical infrastructure in low-lying coastal zones and small island developing states [RFC 1, 2, 3, 4, and 5].
- Risk of food insecurity and the breakdown of food systems linked to warming, drought and precipitation variability particularly in regions that are characterized by poorer populations in urban and rural settings [RFC 2, 3 and 4].
- Risk of severe harm[11] due to inland flooding and the limited coping and adaptive capacities of large urban populations [RFC 2 and 3].
- Risk of loss of rural livelihoods and income of rural residents due to insufficient access to drinking and irrigation water, and reduced agri-cultural productivity, as well as risk of food insecurity, particularly for farmers and pastoralists with minimal capital in semi-arid regions [RFC 2 and 3].
- Systemic risks due to multiple interacting hazards affecting infrastructure in combination with a high dependency of people on critical services (electricity, water supply, health and emergency services) which may break down during extreme events [RFC 2, 3, and 4].
- Risk of loss of marine ecosystems and the services they provide for coastal livelihoods. Biodiversity and coastal ecosystem services important for fishing communities in the tropics and the Arctic are especially at risk due to rising water temperature and the increase of stratification and ocean acidification [RFC 1, 2, 3, 4, and 5].
- Risk of loss of terrestrial ecosystems and the services they provide for terrestrial livelihoods. Biodiversity and terrestrial ecosystem services are important for rural and urban communities globally. These services are at risk due to rising temperatures, changes in precipitation patterns, and extreme weather events. Risks are high for communities whose livelihoods depend on provisioning services [RFC 1, 3, and 4].

- Risk of mortality, morbidity, and other harms during periods of extreme heat, particularly for urban populations of the elderly, infants, people with chronic diseases or compromised immune systems, and expectant mothers. Increasing frequency and intensity of extreme heat (including exposure to the urban heat island effect and air pollution) interacts with an inability of some local organizations that provide health, emergency, and social services to adapt to new risk levels for vulnerable groups [RFC 2 and 3].[12]

Clearly there are many reasons to be concerned about climate change.

Looking at a figure from AR5 – Figure 1.1 here – is helpful for seeing how harmful climate change may be and how our actions will affect the extent of harmful climate impacts. The left side of Figure 1.1 illustrates how a high emissions scenario will lead to a much warmer – and changed – planet than a low emissions scenario. The right side of Figure 1.1, which is often referred to as the "burning embers diagram," helps us visualize what this means in terms of how harmful the related climate impacts will be for all five RFC. Figure 1.1 illustrates that high-emissions scenarios will lead to high risk in all categories by the end of the century, whereas a low emissions scenario carries less, though not negligible, risks in most categories. At first glance it hence seems obvious that climate change is and will be harmful to many people, with more significant and broader impacts the higher our emissions are. The only question is how high global average surface temperatures will climb and what the direct and indirect impacts of this will be. So in the most basic sense it seems climate change is and will be harmful; it is just a matter of how bad it will get. However, how and why is this so?

2 Risk, Uncertainty, and the Complexity of Climate Change's Harmful Effects

Generally something is understood to be harmful when it is bad for someone either by setting back her interests, by physically hurting her, by violating her rights, by impairing or handicapping her, and so on. That is, someone has been harmfully affected when she is made worse off relative to some prior state or benchmark.[13] Intuitively, climate change does and will do just this; it will make people worse off in all kinds of ways through both direct and indirect impacts. Some individuals will be physically injured or killed directly by increased summer temperatures or especially strong storms; others will be negatively impacted by changes in agricultural productivity or the spread of vector-born diseases.

Digging a bit deeper we can see that sometimes when we think about harm we think about the ways in which people can harm each other. We generally think it is wrong to harm others, though we sometimes find unintentional acts of harm to be more permissible than intentional acts of

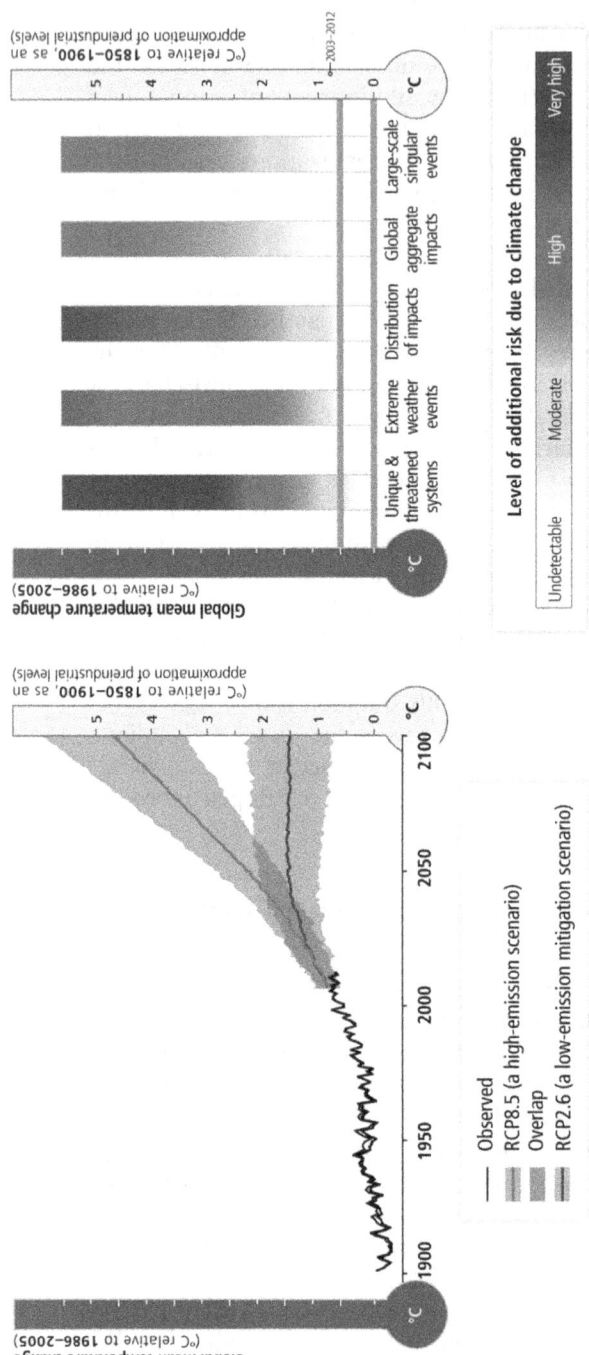

Figure 1.1 Human Interference with the Climate System

Source: Field et al. 2014. Visit www.ipcc.ch/report/graphics/index.php?t=Assessment Reports&r=AR5 - WG2&f=Technical Summary for color image.

Note: A global perspective on climate-related risks. Risks associated with reasons for concern are shown on the right for increasing levels of climate change. The color shading indicates the additional risk due to climate change when a temperature level is reached and then sustained or exceeded. Undetectable risk (white) indicates no associated impacts are detectable and attributable to climate change. Moderate risk (yellow) indicates that associated impacts are both detectable and attributable to climate change with at least medium confidence, also accounting for the other specific criteria for key risks. High risk (red) indicates severe and widespread impacts, also accounting for the other specific criteria for key risks. Purple shows that very high risk is indicated by all specific criteria for key risks. For reference, past and projected global annual average surface temperature is shown on the left [WGI AR5 Figures SPM.1 and SPM.7]. Based on the longest global surface temperature dataset available, the observed change between the average of the period 1850–1900 and of the AR5 reference period (1986–2005) is 0.61°C (5–95% confidence interval: 0.55 to 0.67°C) [WGI AR5 SPM, 2.4], which is used here as an approximation of the change in global mean surface temperature since preindustrial times, referred to as the period before 1750 [WGI and WGII AR5 glossaries].

harm. Other times we do not distinguish between a person harming another person and a tornado harming a person. When we focus on harmful outcomes we do not always distinguish between outcomes that resulted from the intentional acts of humans (including their unintended consequences) and those that resulted from non-anthropogenic causes. At the same time, while both the harm caused by a tornado and a terrorist attack are bad, our moral reaction to these bads is different. It matters whether a harmful outcome is the result of a wrongful as opposed to merely a tragic act or event. When we think about harm as relational between people, ethical questions, particularly of responsibility, come to the forefront. However, when we think of harmful outcomes that are not owing to the actions of humans we do not attribute ethical responsibility to whatever caused the harm in the same way. We may think that a tornado and terrorist attack are both devastating and that we ought to do something to alleviate the harmful effects of both of these events, but whereas we do not think the tornado does anything wrong when it harmfully affects people (in fact we cannot even make sense of what this would mean), the terrorist almost certainly harms his victims and in so doing wrongs them. We usually associate human-caused or anthropogenic harms with wrongdoing, whereas we do not associate non-anthropogenic harms with any wrongdoing even though these can have similarly tragic outcomes.

Climate change is anthropogenic; humans are causing it. However, climate change itself does not directly on its own lead to harmful outcomes because of the complexity of the causal forces at work. The tricky part here is that on the one hand a specific climate impact, such as an especially harmful hurricane, seems like it would fall into the non-anthropogenic category of harming since it is a physical phenomenon just like a tornado; on the other hand, if a specific harmful climatic impact is owing at least in part to the acts of humans, then it seems that the harmful outcomes should fall into the category of anthropogenic harm. How do we distinguish anthropogenic vs. non-anthropogenic harms in the context of climate change? Do actions that contribute to climate change constitute acts of harming because of climate change's harmful effects? How should we understand the harmfulness of climate impacts?

The causal complexity of climate change makes answering these questions extremely difficult. The causal chain between GHG emissions and a specific harmful event such as a specific hurricane, drought, or flood is difficult to map out. This in turn makes it difficult to attribute specific harmful outcomes to climate change and/or identify any specific climate impacts as the result of anthropogenic harm. There is mounting evidence, for example, that the strength and frequency of hurricanes is increasing because of anthropogenic changes to the global climate,[14] but the causal complexity involved (e.g., what differentiates climate from weather) is such that we cannot attribute a specific hurricane or its intensity – let alone a specific harmful outcome such as the death of an individual – to climate change. There certainly *is* evidence that anthropogenic climatic changes are causally contributing to harmful outcomes, but our understanding of climate change is limited both by the complexity of the causal relationships involved and our limited (though rapidly expanding) understanding of this system and how it is responding to human influence. This makes it difficult if not impossible to attribute to climate change a specific harmful effect of a specific climate impact.

Our understanding is further complicated by the fact that climatic changes themselves are not necessarily harmful on their own. Myriad other factors determine the nature and extent of the harmfulness of any particular climatic impact. Even direct impacts, such as deaths caused by extreme climatic events, are influenced by other factors such as the nature of local emergency evacuation procedures, resources, and infrastructure. Hurricane Katrina, for example, was no doubt harmful in part because of failures to respond appropriately at federal, local, and individual levels. So we cannot simply ask if climate change harmed the victims of Hurricane Katrina. We must also ask if those responsible for emergency preparedness and planning harmed or failed to protect them; if some victims harmed or failed to protect themselves because of their choices (e.g., not to evacuate); and so on. Because the harmful outcomes of an event like Hurricane Katrina often have many complex and interconnected causes, it is difficult to answer the seemingly simple question of who or what harmfully affected its victims.

Framing this in terms of the different contributors to climate risk discussed above, we can see that it was not simply the hurricane (which may or may not be a climate impact) but also the fact that the people living in New Orleans and the surrounding area were exposed to the hurricane's harmful effects in virtue of living where they did combined with their social vulnerability, which was caused by a range of factors, that put them at risk and led to their being so harmfully effected. Given that climate change impacts will take many forms, that often we will not be able to attribute a specific event to climate change, and that vulnerability depends on many other factors (e.g., local infrastructure, wealth, and resources), it is extremely difficult, if not impossible, to attribute specific harmful outcomes to climate change. This is the case despite the fact that it is not at all difficult to back up the claim that anthropogenic forces will causally contribute to harmful outcomes via climate change.

Just this much makes it clear that precaution is an appropriate concept to apply to climate change because risk and uncertainty pervade our understanding of it. Yet these concepts – risk and uncertainty – are understood in different ways in different contexts. Economists are usually very precise in defining risks, distinguishing risk from uncertainty: risk is defined as randomness with knowable probabilities, while uncertainty is defined as randomness with unknowable probabilities.[15] The distinction between risk and uncertainty may also be described in terms of measurable and immeasurable uncertainties. Risks involve measurable uncertainties, whereas what we might call genuine uncertainties involve immeasurable or unmeasured uncertainties. While it is tempting to think that climate change involves risk because we are able to quantify and model future changes on various scenarios,[16] this is mistaken because of genuine uncertainties that continue to pervade our understanding of climate change. We certainly *can* model climate change, yet the existence of genuine uncertainties makes any modeled results inherit some of this uncertainty. So while climate change *looks like* it technically involves risk, it is important to recognize that genuine uncertainties continue to pervade our knowledge, hence undermining it fitting this model.

So when the IPCC identifies the above key risks posed by climate change they certainly do not mean to imply that all of such risks can be precisely quantified. The IPCC defines risk such that risks involve:

> The potential for consequences where something of value is at stake and where the outcome is uncertain, recognizing the diversity of values. Risk is often represented as probability of occurrence of hazardous events or trends multiplied by the impacts if these events or trends occur. Risk = (Probability of Events or Trends) × Consequences.[17]

So while the latter part of this definition may seem to track the economic definition of risks, the first part captures the broader sense of "risk" that the IPCC uses when discussing key risks and reasons for concern about climate change.[18] In fact, the IPCC goes on to say that "[c]limate change is not a risk per se; rather climate changes and related hazards interact with the evolving vulnerability and exposure of systems and therewith determine the changing level of risk."[19] Climate impacts alone are not harmful. What makes climate change harmful are the ways in which climate impacts interact or intersect with people's exposure and vulnerability to these impacts. Climate risk then is created by the combination of climate impacts and people's physical exposure and social vulnerability.[20] This is the much broader notion of risk I gesture towards in the title of this book.

In saying climate change creates a climate of risk I therefore mean at least two things. First, climate change threatens to have diverse harmful outcomes in which many things of human value will be adversely impacted. Second, there is a significant degree of uncertainty about both just how bad these impacts will be, and when and where specific impacts will occur. So while we can assess the risks climate change poses using the formula suggested above, uncertainties pervade our understanding of the probability of events and the likely consequences of climate change. Climate change is risky in ways that do not neatly fit the economic definition of risk because our understanding of possible climate impacts falls somewhere in between knowable and unknowable probabilities. Some risks are well understood though uncertainty about them persists, while others are much more uncertain or even unknown, though theoretically possible (e.g., so-called "climate surprises"), at this point. The complexity of the climate system and our ever-expanding understanding of it mean that there are always significant uncertainties involved in our assessment of it.[21]

To address the issue of uncertainty, the IPCC uses a system of likelihood and confidence statements to quantify their judgments as to the likelihood of various outcomes and their confidence in their results. In AR4 the IPCC explicitly admitted that there is uncertainty in our understanding of possible climate outcomes. That report stated that, "estimating uncertainties is intrinsically about describing the limits to knowledge and for this reason involves expert judgment about the state of that knowledge."[22] Because of this, and other uncertainties inherent to understanding the global climate system, "[t]he uncertainty guidance for the Fourth Assessment Report draws, for the first time, a careful distinction between levels of confidence in scientific understanding and the likelihoods of specific results."[23]

The IPCC's 2013–14 AR5 continues to differentiate between confidence and likelihood statements, though it notes that these concepts have evolved since they first appeared in AR4. In AR5 confidence and likelihood statements are understood such that: "Confidence in the validity of a finding, based on the type, amount, quality, and consistence of evidence (e.g., data, mechanistic understanding, theory, models, expert judgment) and the degree of agreement. Confidence is expressed qualitatively."[24] The five qualifiers of confidence are qualitative and range through very low, low, medium, high, and very high. Likelihood statements, on the other hand, reflect, "Quantified measures of uncertainty in a finding expressed probabilistically (based on statistical analysis of observations or model results, or expert judgment."[25] The IPCC uses the following system for ranking likelihood statements:

Virtually certain	99–100% probability
Extremely likely	95–100% probability
Very likely	90–100% probability
Likely	66–100% probability
More likely than not	50–100% probability
About as likely as not	33–66% probability
Unlikely	0–33% probability
Very unlikely	0–10% probability
Extremely unlikely	0–5% probability
Exceptionally unlikely	0–1% probability[26]

Distinguishing confidence from likelihood enables the IPCC to communicate different types of uncertainties in our understanding of climate change. Confidence assessments enable IPCC scientists to indicate their expert scientific judgment, even when this cannot be precisely quantified, whereas likelihood statements reflect quantitative uncertainties that, for example, are reflected in complex climate modeling results. So when the IPCC says, for example, "[i]t is *extremely likely* that human influence has been the dominant cause of the observed warming since the mid-20th century,"[27] we can understand that there is quantitative evidence supporting the conclusion that there is at least a 95% probability that observed climate change is large anthropogenic.[28]

The IPCC therefore helps us understand just *how well* we understand various aspects of and implications of climate change. Looking back at Figure 1.1 it is clear that while we can be extremely confident *that* climate change is happening, uncertainty still pervades our understanding of just *how bad* things will get given different emissions scenarios. There are two key things to look at in Figure 1.1 that reveal the extent to which uncertainty pervades our understanding of the details here, which is really a gross understatement – as will quickly become clear.

First, note that there is a small amount of uncertainty even in our assessment of observed change. It is noted that the observed change between the average global surface temperature from 1850–1900 and 1986–2005 is 0.61°C, but this is really the midpoint of the 5–95% confidence interval that observed warming has been between 0.55°C to 0.67°C. So even our comprehension of observed warming has to be understood in terms of a range rather than a precise figure because of the complexity of the climate system and our ability to aggregate data in the calculation of global mean surface temperature. Simply put, we do not have thermometers on every surface of the planet, so we have to extrapolate from a range of data sources to determine global mean surface temperature. Note that the point here is *not* that there is any question about *whether* global mean surface temperature has increased or whether this increase is due to human influence – both of these facts are very well established by the science; the point, rather, is that studying even our current climate is complex and this means that uncertainty enters into our understanding of it.

Second, note that for the two Representative Concentration Pathways (RCPs) that are illustrated, a shaded area surrounds a dark line. The dark lines denote what is called the ensemble mean, or the average predicted temperature given the RCP, while the shaded area denotes the ±1.64 standard deviation range. This helps us visualize both that the higher emissions scenario will cause significantly more warming but also that with any emissions scenario we can at best predict a range for global mean surface warming at a given point in time. Seeing this next to the burning embers diagram that visualizes the risks associated with increased surface temperatures makes it very clear that the question is not if there will be warming, but how much warming there will be. Higher cumulative emissions will cause greater warming, but there is a significant difference between 3.5°C and 5.5°C of warming in terms of risks. So here again we see that the connection between higher emissions and higher climate risks is not in question. Yet there is significant uncertainty about just what the future will look like with a high emissions scenario both because we do not know precisely how much warming will ensue (and the difference between 3.5°C and 5.5°C is vast) and precisely what climate impacts will occur at what temperature (though we know higher temperatures mean greater risks). It is also worth noting that the graph in Figure 1.1 only goes out to 2100. While this may seem like a sufficient timescale to model, the cumulative emissions scenarios illustrated will impact the climate for thousands of years. Yet the farther into the future we go, the more uncertainty enters into our understanding of precisely what will happen given different emissions scenarios.

Furthermore, while we know a lot about climate change, our understanding of it is continuously emerging. Every year we learn more about the immensely complex climate system, and every year we learn more about just how dire the situation is not just because we continue to emit GHGs but also because climate impacts we thought were mere possibilities are becoming our disturbing reality as our evolving understanding of melting in the Antarctic (discussed in the Introduction) exemplifies. It is because despite all we know climate change continues to surprise us in terms of how quickly and extensively it will harmfully affect us that precaution is an appropriate concept to apply to it. It is no longer reasonable to argue that climate change is not happening. It is largely *because* uncertainties continue to pervade our understanding of climate change that we have strong moral reasons to take much stronger precautionary measures against climate change than even many climate advocates suggest.

What all of this illustrates is that it may be best to understand how climate change is harmful from a big-picture perspective. Climate change is and will be harmful, despite the fact that we may rarely ever be able to claim climate change alone caused a particular harmful outcome. While most events do not have a single cause, climate change is especially complex both because the mechanisms of climate change are complex and its effects are influenced by context. This layered complexity of climate change means that there will often be several causal factors – complex climate factors and external, situational factors – contributing to any given harmful climate change impact. The extent to which climate change will be harmful is still uncertain because: (1) the complexity of the climate system is such that while we can predict with confidence *that* climate change will have harmful effects, uncertainty pervades our understanding of exactly how the climatic system will respond to increased GHG concentrations (e.g., what the particular physical impacts will be); (2) future GHG emissions are unknown; and (3) how harmful physical climate impacts are will depend on the physical, social, and economic context in which they manifest (e.g., whether they occur in a developed vs. a developing country). This significantly complicates the question of whether or not actions that contribute to climate change are acts of harm. Climate change is and will be harmful, but this harmfulness is complex.

3 Harmfulness Across Generations[29]

An additional challenge to understanding how and why climate change is harmful stems from the very concept of "harm" and the challenge of applying this concept over very long time horizons. Whatever we do or do not do to address climate change will significantly shape the world, just as the actions that have led to the situation we are now in shaped our existence. It turns out that the very same actions that affect climate change will also affect which particular people will come to be harmfully affected (or not) by climate change. As Dale Jamieson says, "[c]limate change will remake the world as well as repopulating it. Climate change will produce a world that is radically different from the one that would otherwise have existed."[30] Through our climate-affecting activities and policies we are shaping the future and affecting which individuals will come to exist.

This fact – that our actions shape the future in profound ways – challenges the ways in which we colloquially talk about harming future generations. While it is temping to say that we are harming future generations by failing to address climate change, there is a sense in which this is misleading. Yes, we are harmfully affecting future generations, but technically we are not making the particular individuals who come to exist worse off than *they* would have been. For if we take action and minimize how harmful climate change becomes we will shape the future and contribute to different individuals coming to exist. While it is tempting to say that someone who is negatively impacted by climate change is *harmed* by climate change or by everyone who has emitted GHGs (or unnecessarily emitted GHGs), it is hard to say that she is made worse off, for this begs the question, "worse off than what?" The actions that have contributed and will contribute to climate change have such profound societal implications that whoever is affected by climate change will also have come to exist in part because of the very same set of actions that led to their being harmfully affected. To avoid confusion it is therefore helpful to focus on this concept of harmfulness, rather than harming, in an intergenerational context like climate change.[31]

Joel Feinberg's introduction of the concept of harmful conditions in the context of discussing harmless wrongdoing is helpful here.[32] Feinberg says:

> We can mean by the phrase *harmful condition* a state in which a person is handicapped or impaired, a condition that has adverse effects on his whole network of interests. By a *harmed condition*, on the other hand, we can mean a harmful condition that is the product of an act of harming.[33]

An act can have harmful effects that do not technically harm by creating or promoting harmful conditions. The same kind of effect can be a harmed condition in one instance and a harmful condition in another. For example, Julie could be in a harmed condition after Justin pushes a boulder down a hill injuring her, but if the same boulder injured her after having been loosened by a falling tree branch Julie would be in a harmful condition. The key difference is the nature of the causal relationship between the act or event in question, the effects of these acts or events, and the person or people affected.

The harmful vs. harmed condition distinction helps clarify at least one key difference between contemporary and future effects of actions. An action puts one's contemporary into a harmed condition when it adversely affects her network of interests or otherwise makes her worse off, but no present action can put a distant future person into a harmed condition since no present action can harm distant future people. Put another way, no distant future person will ever be able rightly to claim she exists in a harmed condition because of acts performed before her conception if these acts helped determine her existence. Nonetheless, future people can be harmfully affected by acts that contributed to their existence insofar as from their perspective (in time) they can be handicapped, impaired, or have their interests negatively impacted by the choices and action of individuals who lived before them.

From her perspective, such a person may think of herself as having been made worse off insofar as she has been negatively impacted, but this worsening is not the result of an act of harm since the act that caused the relative worsening also contributed to her existence. In order for an act to be an act of harm it must, from the perspective of the actor, make whomever it affects worse off. The point of comparison for a harmful condition, however, is the person experiencing the condition. Someone is in a harmful condition whenever, from her own perspective (in time), she has been made worse off, where the causes of her condition were not acts of harming. So whereas to determine whether Justin harmed Julie we look at the effects of Justin's action on Julie, to determine if the tree branch-loosened boulder harmfully affected Julie (i.e., put her into a harmful condition), once we establish no action caused her harm, we look at the effects of the boulder on Julie from her own perspective.

This distinction between harmful and harmed conditions can help us make sense of the difference between a harmful terrorist attack and a harmful tornado. Those harmfully affected by a terrorist attack are in a harmed condition, since terrorists acted in ways that caused harm. As long as the tornado is directly causally responsible for the relevant negative impacts, those affected by the tornado, however, are in a harmful condition. A tornado cannot harm because it cannot intentionally act. The tornado victims have been made worse off from their own perspective, but no *act* made them worse off, hence no one harmed them. As was hinted above, however, we can do things that make tornadoes more or less harmful. For example, advance-warning systems can cause less people to be harmfully affected by tornadoes whereas lax building codes may make tornadoes more harmful. As in the case of climate change, tornado victims are affected by a complex set of causal forces. While we may be able to attribute some harm directly to the acts of humans in such a case, at the very least we can say such victims are in a harmful condition, though this may simultaneously be a harmed condition as well.

At least with respect to climate change, those harmfully affected by climate impacts are and will be in harmful conditions. Of course other causal factors such as building codes or the acts of contemporary individuals may affect precisely how harmful climate change is, but as *a* causal contributor to harmful effects, climate-affecting activities can only harmfully affect – not harm – future people. Since the effects of climate change are the result of temporally diffuse actions, no single climate-affecting action may cause any specific individual to be in a harmed condition. The challenge, given all of this, is to identify and articulate what moral obligations we have with respect to climate change. The implication is not that ethics is not relevant here but rather that we have to be careful to acknowledge the complex way in which climate change is harmful when sorting out what we should do to address it.

4 Conclusion

How is climate change harmful? The answer to this question is simultaneously simple and complex because climate change is a global, intergenerational phenomenon that defies many of our colloquial notions. Climate change is and will harmfully affect many people, though what actions we take to address the causes and effects of climate change will impact both the extent of this harmfulness and who will be harmfully affected. This is because we can affect the identity of future people as well as the exposure and vulnerability to climate risk of whoever comes to exist. Climate change hence poses extremely complex intergenerational risks. The rest of the book aims to help us understand what we should do in the face of the diverse risks climate change poses, and how we should approach decision making about climate policy in a way that recognizes and addresses this complexity. Precaution turns out to be a useful guiding concept for climate policy because it addresses concerns about possible harmful outcomes even in the face of uncertainty.

Notes

1 For a discussion of why the intergenerational nature of climate change is amongst its most morally challenging features, see Gardiner 2011, 2006.
2 IPCC 2013.
3 IPCC 2013, 11.
4 IPCC 2013, 11.
5 Archer 2008; Ciais et al. 2013, 472–73.
6 UNFCCC 1992.

 7 Oppenheimer et al. 2014.
 8 IPCC 2007, 284–89.
 9 Oppenheimer et al. 2014, 1049.
10 Oppenheimer et al. 2014, 1048.
11 Note that what is meant by "harm" here is something like "damage."
12 Oppenheimer et al. 2014, 1042.
13 This is not to deny that animals, plants, or ecosystems can be harmfully affected. I focus on the ways in which humans can be harmfully affected, though parallel inquiries could explore the ways in which nonhuman animals, plants, ecosystems, etc. may be harmfully affected.
14 Rhein et al. 2013.
15 Knight 2002.
16 One example of this is when it is (mistakenly) claimed that we do not need the precautionary principle to guide climate policy because climate change is technically a risk. See Vinuales 2010.
17 Oppenheimer et al. 2014, 1048.
18 For a discussion of the role of risks in AR5, see also Mach et al. 2016.
19 Oppenheimer et al. 2014, 1050.
20 For an excellent discussion of why it is important to pay attention to the sources of vulnerability in assessing the moral significance of climate risk, see Blomfield 2015.
21 See Chapter 4 for a detailed discussion of this issue. I there demonstrate how a range of uncertainties pervades even the most rigorous assessments of climate change and why this is ethically significant. It is also important to note, however, that the precautionary view I defend in subsequent chapters suggests that even if we were 100% certain of climate catastrophe, precaution would still be a useful guide since in this case it guides us to avoid catastrophic outcomes, whether we are uncertain about them or not.
22 Solomon et al. 2007, 22.
23 IPCC 2013, 22.
24 Stocker et al. 2013, 36.
25 Stocker et al. 2013, 36.
26 Stocker et al. 2013, 36.
27 IPCC 2013, 17.
28 For a discussion of uncertainty in our understanding and interpretation of climate science, see Oreskes 2010.
29 See the Appendix for further discussion of the topics covered in this section. I there offer an in-depth discussion of the non-identity problem, a philosophical puzzle underlying this brief discussion, and offer an expanded argument for thinking of climate impacts in terms of their harmfulness rather than as harming individuals per se.
30 Jamieson 2014, 167.
31 The Appendix fleshes out the arguments underlying this discussion in greater detail.
32 Feinberg 1990.
33 Feinberg 1990, 26.

References

Archer, David. 2008. *The Long Thaw: How Humans are Changing the Next 100,000 Years of Earth's Climate.* Princeton: Princeton University Press.

Blomfield, Megan. 2015. "Climate Change and the Moral Significance of Historical Injustice in Natural Resource Governance." In *The Ethics of Climate Governance*, ed. Aaron Maltais and Catriona McKinnon. Rowman & Littlefield.

Ciais, P., C. Sabine, G. Bala, L. Bopp, V. Brovkin, J. Canadell, A. Chhabra, et al. 2013. "2013: Carbon and Other Biogeochemical Cycles." *Climate Change 2013: The Physical Science Basis. Contribution of Working Group I to the Fifth Assessment Report of the Intergovernmental Panel on Climate Change*, 465–570. doi:10.1017/CBO9781107415324.015.

Feinberg, Joel. 1990. *Harmless Wrongdoing: The Moral Limits of the Criminal Law*. New York: Oxford University Press.

Field, C.B., V.R. Barros, K.J. Mach, M.D. Mastrandrea, M. van Aalst, W.N. Adger, D.J. Arent, et al. 2014. "Technical Summary." In *Climate Change 2014: Impacts, Adaptation, and Vulnerability. Part A: Global and Sectoral Aspects. Contribution of Working Group II to the Fifth Assessment Report of the Intergovernmental Panel on Climate Change*, ed. C.B. Field, V.R. Barros, D.J. Dokken, K.J. Mach, M.D. Mastrandrea, T.E. Bilir, M. Chatterjee, et al., 35–94. Cambridge: Cambridge University Press.

Gardiner, Stephen M. 2006. "A Perfect Moral Storm: Climate Change, Intergenerational Ethics and the Problem of Moral Corruption." *Environmental Values* 15: 397–413.

Gardiner, Stephen M. 2011. *A Perfect Moral Storm: The Ethical Tragedy of Climate Change*. Oxford: Oxford University Press.

IPCC. 2007. "Climate Change 2007: Synthesis Report. Contribution of Working Groups I, II and III to the Fourth Assessment Report of the Intergovernmental Panel on Climate Change." In *IPCC*, ed. Core Writing Team, R.K. Pachauri, and A. Reisinger, 1:1:104. Cambridge University Press.

IPCC. 2013. "Summary for Policymakers." In *Climate Change 2013: The Physical Science Basis. Contribution of Working Group I to the Fifth Assessment Report of the Intergovernmental Panel on Climate Change*, ed. P.M. Midgley, T.F. Stocker, D. Qin, G.-K. Plattner, M. Tignor, S.K. Allen, J. Boschung, A. Nauels, Y. Xia, V. Bex, 1–30. Cambridge: Cambridge University Press. doi:10.1017/CBO9781107415324.004.

Jamieson, Dale. 2014. *Reason in a Dark Time: Why the Struggle Against Climate Change Failed – and What it Means for Our Future*. Oxford: Oxford University Press.

Knight, Frank Hyneman. 2002. *Risk, Uncertainty and Profit*. Beard Books.

Mach, Katharine J., Michael D. Mastrandrea, T. Eren Bilir, and Christopher B. Field. 2016. "Understanding and Responding to Danger from Climate Change: The Role of Key Risks in the IPCC AR5." *Climatic Change* 136(3–4): 427–444. doi:10.1007/s10584-016-1645-x.

Oppenheimer, M., M. Campos, R. Warren, J. Birkmann, G. Luber, B. O'Neill, and K. Takahashi. 2014. "Emergent Risks and Key Vulnerabilities." In *Climate Change 2014: Impacts, Adaptation, and Vulnerability. Part A: Global and Sectoral Aspects. Contribution of Working Group II to the Fifth Assessment Report of the Intergovernmental Panel on Climate Change*, ed. C.B. Field, V.R. Barros, D.J. Dokken, K.J. Mach, M.D. Mastrandrea, T.E. Bilir, M. Chatterjee, et al., 1039–1099. Cambridge: Cambridge University Press.

Oreskes, Naomi and Erik M. Conway. 2010. *Merchants of Doubt: How a Handful of Scientists Obscured the Truth on Issues from Tobacco Smoke to Global Warming*. Bloomsbury Press.

Rhein, M., S.R. Rintoul, S. Aoki, E. Campos, D. Chambers, R.A. Feely, S. Gulev, et al. 2013. "Observations: Ocean." In *Climate Change 2013: The Physical Science Basis. Contribution of Working Group I to the Fifth Assessment Report of the Intergovernmental Panel on Climate Change*, ed. T.F. Stocker, D. Qin, G.-K. Plattner, M. Tignor, S.K. Allen, J. Boschung, A. Nauels, Y. Xia, V. Bex, and P.M. Midgley, 255–316. Cambridge: Cambridge University Press. doi:10.1017/CBO9781107415324.010.

Solomon, S., D. Qin, M. Manning, R.B. Alley, T. Berntsen, N.L. Bindoff, Z. Chen, et al. 2007. "Technical Summary." In *Climate Change 2007: The Physical Science Basis Contribution of Working Group I to the Fourth Assessment Report of the Intergovernmental Panel on Climate Change*, ed. S. Solomon, D. Qin, M. Manning, Z. Chen, M. Marquis, K.B. Averyt, M. Tignor, and H.L. Miller, 23–78. Cambridge: Cambridge University Press. doi:10.1007/s10894-008-9162-1.

Stocker, T.F., D. Qin, G.-K. Plattner, L.V. Alexander, S.K. Allen, N.L. Bindoff, F.-M. Bréon, et al. 2013. "Technical Summary." In *Climate Change 2013: The Physical Science Basis. Contribution of Working Group I to the Fifth Assessment Report of the Intergovernmental Panel on Climate Change*, ed. T.F. Stocker, D. Qin, G.-K. Plattner, M. Tignor, S.K. Allen, J. Boschung, A. Nauels, et al., 33–115. Cambridge: Cambridge University Press. doi:10.1017/ CBO9781107415324.005.

UNFCCC. 1992. *United Nations Framework Convention on Climate Change*. http:// unfccc.int/resource/docs/convkp/conveng.pdf.

Vinuales, Jorge E. 2010. "Legal Techniques for Dealing with Scientific Uncertainty in Environmental Law." *Vanderbilt Journal of Transnational Law* 43(437): 437–503.

2   Making Sense of Precaution

The last chapter established that while it is certain *that* climate change will happen, significant uncertainty pervades our understanding of exactly *what* will happen *when*. This is precisely why precaution is an appropriate concept to apply to climate change. We need guidance in the face of questions such as: How much should we be willing to risk when it comes to climate change? What does morality demand of us in this case? It is okay to push for a policy that limits the odds of global average surface temperatures rising 2°C above preindustrial levels to 50%? Would it be morally acceptable to choose a policy pathway that comes with a 5% risk of going above 6°C? How do we know what 2°C or 6°C will even look like? How much certainty should we have about how harmful these scenarios would be to make morally appropriate choices? At first glance these are the kinds of questions the precautionary principle should be able to help us answer, but what exactly is the precautionary principle, and how can it help guide us to specific climate policies?

In this chapter I argue that in fact there is no single precautionary principle. "The precautionary principle" has simply come to stand for too many different things, and some of the core values at its heart are deeply incommensurable. Nonetheless, precaution is a useful concept in many contexts and as such should not be abandoned. Instead a range of precautionary values may be captured in diverse precautionary principles, precautionary approaches, and precautionary decision-making frameworks. Some authors, in fact, have already begun to move away from talk of "the" precautionary principle, at least implicitly recognizing that there is no one precautionary principle.[1] Russell Powell goes so far as to say:

> It is somewhat misleading to refer to *the* precautionary principle, since the doctrine enjoys no canonical formulation. Instead, it amounts to a largely disconnected constellation of legal, political, and academic articulations that fall within the rubric of what might be called the *precautionary approach*.[2]

Nonetheless, Powell continues to discuss "the" precautionary principle and its many formulations, seemingly just because this is the norm in the literature. I, however, urge us to latch onto the idea that there is no referent to "the precautionary principle." Precaution does have a powerful role to play in environmental thought and policy, but there is no one single principle that is *the* precautionary principle.

To argue for this view I start by exploring what we might call the myth of "the" precautionary principle. I briefly point out that we cannot be required to take precautionary measures against any and all threats of harm, lest we be led into a precautionary paradox. I then consider what is meant when authors talk of "versions" or "formulations" of the precautionary principle. I suggest that what must be implied is a kind of unified family tree view of the precautionary principle. Next I address a significant tension that arises within the family tree view from the idea that the precautionary principle addresses threats of harm to both human health and the environment. As valuing human health and the environment often pull us in different directions, it is incoherent to value both simultaneously without any guidance about what to do when they conflict. I then consider the idea that the precautionary principle should be understood as an approach to risk management such that it is more of a decision-making procedure or framework than a moral principle. Here again I find it hard to understand where "the" precautionary principle is to be found in this view.

In the second section I then take a close look at Daniel Steel's attempt to unify the precautionary principle, arguing that while many features of his view are attractive, it is still very unclear what "the" precautionary principle is in his view, at least in part because his view inherits the aforementioned tension. This leads me, in the third section, to draw on those who understand the precautionary principle as a banner for a new way of thinking about risk management in the face of uncertainty to argue that there is no single precautionary principle. Rather, what unifies work on the so-called precautionary principle is quite simply the concept of precaution. More strongly put, we ought to abandon the idea that "the" precautionary principle can or should be unified by anything more than this loose notion. The way forward, I go on to argue, is to recognize that there is a range of loosely connected precautionary principles, precautionary decision-making frameworks, and precautionary approaches, each of which needs to be carefully identified and defended in light of the many insights the literature on precaution provides.

1 The Myth of "the" Precautionary Principle

The precautionary principle has been proposed as a possible guide to addressing all sorts of threats of harm in all sorts of contexts. It is often associated with the intuition that it is better to be safe than sorry; when faced with a threat of harm we should take precautionary measures even if we are uncertain about its likelihood or extent. Alternatively the precautionary principle is sometimes associated with the idea that we ought to act in advance of scientific certainty to protect human health and/or the environment. The authors of a paper on the precautionary principle in contemporary environmental policy express this another way when they say, "[a]t its core lies the intuitively simple idea that decision makers should act in advance of scientific certainty to protect the environment (and with it, the well-being of future generations) from incurring harm."[3] This section explores different interpretations of the precautionary principle that, together, ground the argument that there is no single precautionary principle or way to unify all that has come to be associated with this concept.

1.1 Precautionary Paralysis

Whatever the precautionary principle is, it must limit the demandingness of precaution so that it does not lead us down a precautionary rabbit hole. We face threats of harm all the time and there are many different kinds of precautionary measures we can take against many of these threats, but this cannot mean that the precautionary principle applies to all aspects of our lives or that we should take every conceivable precautionary measure against all such threats. Certainly there is a sense in which we might think of the precautionary principle as a guiding rule of thumb we appeal to all the time. Whenever we wear our seatbelt, wash our hands before eating, or carry first aid supplies while hiking we might understand ourselves to be following a kind of prudential precautionary principle that captures the "better safe than sorry" intuition many of us feel compelled to follow. Yet one of the few things on which diverse authors writing on the precautionary principle agree is that this principle cannot apply to *all* threats of harm, nor can it require that we take all available precautionary measures against any given threat, lest it lead us into a precautionary paradox.[4] For if we were required to take all available precautionary measures against every and all threats of harm we would quickly become paralyzed, unable to act at all since virtually all actions carry with them a set of risks, however minute, and against each of these risks we could take any number of precautionary actions. Part of the worry here is that any precautionary action suggested by the precautionary principle will have its own risks such that the precautionary principle would prohibit the very action it seems to recommend. What is often called the paralysis objection or precautionary paradox stems from just this: risks abound on all sides.

1.2 The Family Tree View of the Precautionary Principle

While the precautionary principle cannot demand that we take precautionary measures against any and all risks, many implicitly conceive of it as a limited family tree wherein there are many versions of the precautionary principle.

This kind of view suggests that the precautionary principle takes different forms in different contexts. Carl Cranor, for example, has moved from talking about "the precautionary principle" as if it were a single principle, to recognizing that there are actually many versions of it.[5] In the philosophical literature authors such as Stephen Gardiner, Darrel Moellendorf, Catriona McKinnon, and Alan Randall each defend what they take to be a strong or particularly plausible version of the precautionary principle.[6] Two versions of the precautionary principle from environmental policy that are commonly cited come out of the United Nations Agenda 21, which was adopted at Rio in 1992, and an academic conference on precaution, sometimes called the Wingspread Precautionary Principle. These are as follows.

The Rio Declaration on Environment and Development (1992)

> In order to protect the environment, the precautionary approach shall be widely applied by states according to their capabilities. Where there are threats of serious or irreversible damage, lack of full scientific certainty shall not be used as a reason for postponing cost-effective measures to prevent environmental degradation.[7]

Wingspread Precautionary Principle (1998)

> When an activity raises threats of harm to human health or the environment, precautionary measures should be taken even if some cause and effect relationships are not fully established scientifically.[8]

While these are now canonical examples of the precautionary principle in the environmental realm, the precautionary principle is appealed to in a wide range of settings from bioethics to war theory. For example, US President George W. Bush's expression of concern about weapons of mass destruction, which he used to justify an invasion of Iraq, was commonly thought to be an application of the precautionary principle.[9] Appeals to the precautionary principle have also been used publicly to justify proposals to ban BPA (bisphenol A) in water bottles.[10] Depending on the author and context, the family tree comprising the precautionary principle hence can be very broad or more narrowly confined to a particular realm (e.g., environmental policy).

In the academic literature it is common to distinguish between strong and weak versions or formulations of the precautionary principle. As Jonathan Hughes says:

> The concept of strength and weakness, as applied to formulations of the precautionary principle, is, therefore, multi-dimensional: formulations can vary in strength not only in the respect of the action they recommend but in respect of the seriousness of potential harm and level of evidence required to trigger that action.[11]

Versions of the precautionary principle are also sometimes distinguished as being either normative or epistemic, where the former make moral claims about when precaution is warranted and the latter focus more on addressing what to do in the face of scientific uncertainty.[12] These examples illustrate how the core precautionary intuitions are expressed in different ways in different applications or versions of the precautionary principle.

While the implication of talk of "versions" or "formulations" of the precautionary principle is that there must be something tying all of these versions together, there is disagreement in the literature as to what features versions of the precautionary principle do or should share and hence how unified the family tree is that comprises the precautionary principle. To see this it is helpful to start by looking at Neil Manson's analysis of the common logical structure that all formulations of the precautionary principle share. This is a good example of a structural attempt to codify what counts as a version of the precautionary principle.[13] Manson generally conceives of the precautionary principle such that it tells us to regulate potentially harmful activities when our scientific knowledge of these activities is "significantly incomplete."[14]

At the core of Manson's analysis is the idea "that for a given *activity* that may have a given *effect* on the environment, the precautionary principle is supposed to indicate a *remedy*."[15] According to Manson's analysis, every formulation of the principle has a three-part structure. First, there is a suggested damage condition, which specifies some foreseeable and harmful effect that some activity might have such as "serious," "catastrophic," or "irreversible." Second, there is a knowledge condition, which "specifies the status of knowledge regarding the causal connections" between the activity in question and the harmful effect such as "possible," "reasonable to think," or "not proven with certainty that it is not the case."[16] Finally, there is a suggested remedy, which may consist of anything from cessation of the potentially harmful activity, to funding for research on alternatives, to regulation. This three-part structure of the precautionary principle has also been called the tripod.[17]

For Manson, anything that has the right structure counts as a version of the precautionary principle. So, "[w]hich formulation of the precautionary principle is most favorable to the pro-precaution side in any given situation will depend on the particular political and legal context in which that formulation is to be employed."[18] This leaves little more than a simple structure holding together versions of the precautionary principle in its family tree.

Another challenge to this view is that many versions of the precautionary principle are problematically unclear. Derek Turner and I draw on Manson's analysis to argue that that many versions of the precautionary principle are often unclear in five key ways.[19] First, the precautionary principle is often formulated such that it is unclear who must take responsibility for and bear the cost of precaution. Second, the precautionary principle is unclear and even internally contradictory when it applies simultaneously to threats of harm to human health and the environment (I will expand on this below). Third, it is often unclear what threats of harm a particular version of the precautionary principle addresses. Fourth, what count as precautionary measures are often left unspecified in versions of the precautionary principle. Fifth, it is often unclear how much precaution is required by a given version of the precautionary principle. The canonical Wingspread formulation of the precautionary principle offered above, for example, is unclear in all of these ways (and anchors our discussion of these un-clarities). This is problematic if versions of the precautionary principle are supposed to be action guiding and also further calls into question what is holding the family tree of the precautionary principle together.

Of course, any given version of the precautionary principle can overcome these concerns. James E. Hickey Jr. and Vern R. Walker, for example, propose a set of criteria for formulating the precautionary principle with the aim of removing the uncertainty often associated with it.[20] They argue that we should think of the precautionary principle as an evolving rule that is responsive to increased knowledge and understanding, evolving norms of international law, and scientific progress.[21] Towards this end, they suggest every articulation of the precautionary principle in international law should contain:

i   a reasonably precise statement of the desired environmental goal and the environmental condition that justifies invoking the precautionary principle;
ii   an identification of the jurisdictional scope of the agreed precautionary obligations under the principle;
iii a specification of those human activities for which precautionary measures are required; and,
iv   a clear statement of the precautionary measures that must be undertaken before engaging in a covered activity.[22]

These criteria are meant as constraints upon the content of the precautionary principle's different formulations. Hickey and Walker argue that it must be clear exactly when the precautionary principle requires concrete efforts to achieve a clearly defined environmental goal if it is to be used in effective environmental policy. They think any version of the precautionary principle must be articulated so that it connects human activities with a specific threat of harm where it is clear what activities the precautionary principle is to regulate and when precautionary measures are to be triggered. They add that a "required degree of confidence in the scientific information" about the likely effects must be specified, implicitly addressing the question of what standard of evidence should be required to activate the precautionary principle.[23]

While these are all helpful suggestions insofar as they may lead to clearer versions of the precautionary principle, it does not help us understand what "the" precautionary principle itself is. Even Per Sandin, who has also sought to create more unity amongst versions of the precautionary principle by requiring that we be very explicit and precise when presenting a particular precautionary principle, especially with respect to intentionality, uncertainty, and epistemic reasonableness, admits that "a universal precautionary principle is difficult to conceive."[24] Turner and I had hoped that our analysis would lead us to be able to formulate the precautionary principle in a way that addressed our worries about its lack of clarity, but we ended up concluding that much like the concept of sustainability, the plausibility of the precautionary principle depended on its vagueness. Yet its vagueness is also a significant weakness. We point out that no one "has been sufficiently clear about what kind of a principle it is supposed to be."[25] The lurking questions are: What are all these versions of the precautionary principle versions *of*? What is *the* precautionary principle? Recognizing the diversity of versions or formulations of the precautionary principle reveals that there is no one coherent thing that is the precautionary principle. If the only way to make sense of talk of versions of the precautionary principle is as a big family tree, what an exploration of that tree reveals is that there is nothing holding it all together, nothing unifying the diverse versions of "the" precautionary principle.

1.3 Protecting the Environment vs. Protecting Human Health

The most significant challenge to a unified precautionary principle, especially in environmental contexts, stems from the fact that the precautionary principle has come to be associated both with environmental protection and the protection of human health because these ends sometimes conflict. Of course there is a way in which environmental protection and the protection of human health can be compatible: protecting the environment is often good for human health and (to a lesser extent) vice versa. However, when the precautionary principle is understood to demand environmental protection for the good of the environment itself (rather than as instrumentally valuable to us, humans), we run into a problem that is very familiar in environmental ethics. Valuing the environment for its own sake is at odds with prioritizing the value of human life. We cannot always simultaneously act so as to protect the good of the environment and human health or well-being. This has resulted not only in different versions of the precautionary principle being incommensurable with each other but also in some versions of the precautionary principle being internally incoherent.

Take the Wingspread formulation of the precautionary principle provided above and note that it addresses threats of harm to human health *or* the environment. What is problematic here is that precautionary measures aimed at protecting human health will sometimes conflict with those aimed at protecting the environment. Think of how we might respond to the threats of harm posed by decreased snowpack in the American west.[26] If our aim is to protect human health we will probably think of this as a water management issue and try to find ways to collect and store more water for humans to use throughout the dry summer months, which would entail removing even more water from a quickly drying system. However, if our aim is to mitigate the environmental impacts of reduced snowpack for local plants, animals, and ecosystems we might decide to do just the opposite, ensuring what little water is still present in the system stays in the system. Of course, often what we should do to protect the environment will help protect human health, but the point is that sometimes there is a tension between these ends.

This tension is that anthropocentric value systems take humans and their well-being to be of primary moral concern, whereas non-anthropocentric value systems prioritize the good of the environment itself (either in sum or focus on some aspect of it). So while we could read a version of the precautionary principle like the Wingspread statement and think what it is saying is that we should protect humans and one important way to do this is by protecting the environment on which we depend, there is another way of reading the "or the environment" statement as having a non-anthropocentric meaning, in which case the principle becomes incoherent insofar as it will sometimes recommend incommensurable actions. A single principle cannot coherently tell us to do A and B, when doing B entails not doing A. It is problematic that some versions of the precautionary principle aim to protect human health, others the environment for its own sake, and yet others, like the Wingspread version, both of these, because there is no way to capture these morally incommensurable values in a single, coherent precautionary principle. That is, "the" precautionary principle cannot coherently tell us we ought to protect the environment and human health when what this sometimes entails is a set of conflicting and incommensurable actions.

Of course, there are ways out of the moral dilemma entailed here. We could decide to prioritize either human health or the environment, making it explicit that whenever possible we should seek to protect both but also that when push comes to shove one or the other ought to take priority. That is, we could choose which is our real moral concern, human health or the environment, and frame the precautionary principle solely in terms of this end. Or we could adopt a version of moral pluralism, accepting that sometimes we face genuine moral dilemmas because we really ought to protect human health and the environment for their own sakes. However, even if we take this last route, it would be more coherent to separate out our sometimes incommensurable obligations such that we distinguish between our precautionary obligations to protect human health and those to protect the environment. If "the" precautionary principle tells us both to start storing more spring runoff and to leave the watershed alone, it is incoherent. On the other hand, we may be willing to accept that an environmental precautionary principle tells us we should leave the watershed alone whereas a human health precautionary principle tells us we should start storing more spring runoff and work out what to do about this dilemma. In any case, there is currently no single referent to "the precautionary principle" since there is no agreement as to what its focus is, even if a pluralistic stance is implied by the current lack of agreement. Despite the common implication that there is *a* precautionary principle, the way "the precautionary principle" is talked about suggests otherwise.

Despite this, the precautionary principle is well established in international law.[27] In response to this, Elizabeth Fisher says:

> the great variation in how it has been formulated, interpreted and implemented is not a weakness of the principle. Rather, it attests to how deeply embedded the principle has become in different legal systems and how there are sharp disagreements over the legal issues that it is a response to.[28]

In conceiving of the precautionary principle in legal terms, Fisher argues the precautionary principle, as a principle, should be inherently flexible so that it is more a mode of legal reasoning than a rigid rule. Similarly, Anton Petrenko and Dan McArthur argue, "as any high-level principle, Precautionary Principle has a level of generality that can be resolved through charitable interpretation based on precedent in-line with the spirit of the principle, rather than one based on the abstract semantic probing of the various formulations."[29] The problem is that even a flexible principle should not be able to tell you to do A and not do A at the same time. The fact that "the" precautionary principle, even within a relatively narrow legal context, can imply radically different and even contradictory things (e.g., build levies vs. tear down levies) is deeply problematic. Interpreting "the" precautionary principle as a high-level principle or guide[30] or as a general goal[31] does not get around this problem.

It is interesting that Fisher explicitly says that she uses "precaution" as shorthand for "the precautionary principle"[32] because this seems to imply she sees that it is problematic to talk about "the" precautionary principle when we are not really referring to a distinct principle at all. Fisher is right that we should be flexible in how we respond to threats of harm and select appropriate precautionary measures, but just because "the" precautionary principle has become embedded in international law does not mean that we have to stick to this locution. A more productive way forward is to latch onto the recognition that there is no one precautionary principle and find a new way to talk about what role precaution *should* play in international law.

Marko Ahteensuu and Per Sandin somewhat similarly claim of the precautionary principle that instead of speaking about several principles (at least, apart from judicial studies), it seems to be more fruitful to say that there is only one principle which is formulated (or understood) in various ways. The use of singular and plural may just indicate the fact that the precautionary principle is thought of at different levels of generality.[33]

While there is something to be said for the idea that the precautionary principle captures an idea expressed at many different levels of generality, and Ahteensuu and Sandin may be right to some extent about why talk of "the" precautionary principle has persisted, it is the opposite of fruitful to continue to insist there is but one precautionary principle that is formulated in different ways or at different levels of generality when there is no way to unify all of these versions in a single statement of the overarching principle or describe what unifies the family tree of its many formulations. Daniel Bodansky notes that versions of the precautionary principle "differ along virtually every important dimension."[34] This leads him to suggest that "[g]iving [the precautionary principle] meaning will require hard thinking about what it means to be cautious in particular contexts, rather than continued incantations of the same old formulations."[35] To say that all versions of the precautionary principle and talk of "the" precautionary principle are all part of "the precautionary principle" when there are deep inconsistencies amongst different versions and use of the phrase "the precautionary principle" takes meaning away from the very concept Ahteensuu and Sandin want to rescue.

1.4 Procedural Interpretations of the Precautionary Principle

To make matters even more complex, there is another way to understand the precautionary principle that is entirely different from what we have seen so far. This is to see it as an approach to risk management or guide to rational decision making in the face of uncertainty. In this view, the precautionary principle is understood as a decision-making procedure or framework.[36] This kind of view guides us when we are faced with diverse risks in a way that should lead us toward coherent precautionary action.

One example of this kind of view comes from Indur Goklany, who argues that the only way the precautionary principle can be intelligently understood is as involving six hierarchical criteria for taking a precautionary approach to risk assessment.[37] His interpretation of the precautionary principle is therefore entirely procedural, though this procedure is imbedded with normative content. Goklany's criteria rank threats according to their characteristics and their degrees of certainty.[38] First, the public health criterion prioritizes humans over other species and the environment. Second, the immediacy criterion says that immediate threats should take priority over later threats. Third, the uncertainty criterion asserts that for equivalent threats of harm, more certain threats should take precedence. Fourth, the expectation-value criterion is that for equally certain threats precedence should be given to those with higher expectation value. Fifth, the adaptation criterion suggests threats can be discounted to the extent to which their adverse impacts can be nullified by technological intervention. Finally, the irreversibility criterion prioritizes irreversible or persistent outcomes. Under ideal circumstances, these criteria would be applied one at a time, but Goklany concedes that in practice several criteria often must be applied simultaneously. He further admits that, "[t]here will obviously be instances in which no cut-and-dried answer is readily apparent."[39] Goklany thinks of the precautionary principle not as a principle applying to threats of harm one at a time, but as a precautionary framework for comparing and ranking multiple threats of harm against each other. Hence the precautionary principle does not demand precautionary measures against any and all threats of harm but rather guides us to take precautionary measures against the most significant or pressing threats by following Goklany's criteria.

David Resnik also offers a procedural interpretation of the precautionary principle. While he notes the precautionary principle is a normative principle, he conceives of it as a decision-making principle rather than as a moral principle.[40] He argues the precautionary principle does not tell us what we are morally obligated to do but rather what it would be rational to do. As such, it includes a procedure for determining what is rational. Resnik is motivated to show it is a rational principle of decision making. He argues that the precautionary principle can be understood as being a rational decision-making principle if and only if it applies just to plausible threats and requires taking only reasonable precautionary measures. He says that when determining whether a threat of harm is credible, we should apply epistemic criteria such as coherence, explanatory power, analogy, precedence, precision, and simplicity; when determining whether precautionary measures are reasonable, we should take into account practical considerations such as effectiveness, proportionality, cost-effectiveness, realism, and consistency.

> If it is handled with discretion and care, the [precautionary principle] can be a sensible approach to making important decisions in the midst of scientific uncertainty. Without such clearly defined restraints on its use, the [precautionary principle] can become twisted into a highly politicized, paranoid and irrational rule.[41]

Resnik's interpretation of the precautionary principle is hence rather informal. He sees it as a rough guide for rational decision making.

Procedural interpretations of the precautionary principle have the advantage of building into this principle that its components be clarified on a case-by-case basis, but these interpretations seem to lose the "principle" part of the precautionary principle since procedures are not themselves principles. As Per Sandin notes of procedural interpretations, "'the' precautionary principle soon dissolves beyond recognition."[42] Once again we see that "the" precautionary principle is understood in many different ways, solidifying the argument that there is no single referent to "the" precautionary principle. Certainly there is something to be learned about precautionary decision making from procedural interpretations of the precautionary principle, but such views further diminish the possibility of a single, unified precautionary principle.

2 Against the Possibility of Unifying the Precautionary Principle

Despite all of this, Daniel Steel has recently taken on the challenge of unifying the precautionary principle.[43] He argues that there are three themes or core elements of the precautionary principle which together comprise a unified account: the meta-precautionary principle, the tripod, and proportionality. The meta-precautionary principle captures the precautionary intuition that unites most formulations of the precautionary principle, namely that uncertainty should not be used as a reason for inaction in the face of serious threats to the environment or human health. It is not a decision rule; rather it is a meta-principle insofar as it restricts what decision rules are appropriate in environmental (and human health) decision making to those that are not "susceptible to paralysis by scientific uncertainty."[44]

The tripod is the three-part structure versions of the precautionary principle must have: a knowledge condition, harm condition, and recommended precaution. Of proportionality, Steel says it "is the idea that the aggressiveness of the precaution should correspond to the plausibility and severity of the threat."[45] "Proportionality places constraints on which versions of [the precautionary principle] may be used to justify which precautions in particular contexts."[46] Two principles comprise proportionality: consistency and efficiency. Consistency requires that the precautions recommended by a version of the precautionary principle are not prohibited by that same version. Efficiency requires that less harmful or costly precautions should be preferred. This account is unifying insofar as it "ties together aspects of [the precautionary principle] that are usually treated as separate or even conflicting."[47] Steel argues that together the three components of the precautionary principle function as a decision rule, though no one element of it can fully guide action. In this way, the precautionary principle can be understood as being a procedural requirement, a decision rule, and an epistemic rule all at the same time.

Steel argues that this interpretation avoids both the claim that the precautionary principle is trivial and that it is irrational, what he identifies as a primary dilemma facing the precautionary principle. The precautionary intuition alone is not enough to meaningfully guide action, yet the precautionary principle would be paralyzing if it required that we take precautionary measures against every threat of harm we face. Steel therefore argues that what some have identified as weak versions of the precautionary principle are actually the meta-precautionary principle; they capture the precautionary intuition that underlies most versions of the precautionary principle. Strong versions of the precautionary principle should be understood simply as versions of the precautionary principle, since these capture specific decision rules (and are spelled out in terms of the "tripod"). In Steel's view the precautionary principle is multi-dimensional and he sometimes refers to it as a complex framework.[48]

While Steel thinks that his account should help us understand the precautionary principle as a unified whole, his view inherits many of the problems identified above. Steel says that "the precautionary principle" should be used to refer to action-guiding versions of the precautionary principle, yet this is also the term he uses to describe his view as a whole. This is confusing since "the precautionary principle" is then used in so many ways in his view. How can "the precautionary principle" simultaneously denote his view as a whole as well as any particular version of it? Is "the precautionary principle" a moral principle, as some versions must certainly be in his view? Or is it a decision-making framework, as Steel sometimes suggests? Is the meta-precautionary principle distinct from the precautionary principle or part of it? Steel succeeds in capturing many dimensions of the precautionary principle that have emerged in the literature, yet he does not offer enough clarity about the meaning of the term this view is supposed to unify, namely, "the precautionary principle."

More importantly, Steel does not resolve concerns about the precautionary principle being incoherent if it captures both anthropocentric and non-anthropocentric values. Steel would likely say that his proportionality criterion ensures that an appropriate version of the precautionary principle will be applied in any particular case so that incommensurability is not a problem. This will only work, however, to the extent that cases are described in a way that makes it clear what value (e.g., human life or ecosystem functioning) should be prioritized. We could end up with the same conflicting result when we apply one so-called version of the precautionary principle that applies to environmental protection and another that applies to protecting human health to two sides of the same issue. This problem stems from the fact that Steel is explicit that he is not interested in making a moral argument for precaution; rather he aims to reinterpret the precautionary principle from the perspective of the philosophy of science.[49] Steel acknowledges that choosing a harm condition for an application of the precautionary principle fundamentally involves value judgments,[50] implying that moral justifications for precaution are not inherent to the principle itself but come into play in the selection of versions of the precautionary principle. However, he does not seem to think that this suggests there is any deep inconsistency to his view of the precautionary principle as a whole.

I think this might stem from the fact that Steel's ultimate justification for the precautionary principle is historical; it is driven by a concern that delaying action can and has led to harmful results. Steel says: "The historical argument recommends [the precautionary principle] on the grounds of a history of insufficient precaution in the face of serious environmental and human health hazards."[51] Hence the meta-precautionary principle is core to the view. Steel argues:

> In the case of the historical argument, the errors in question involve harms to human health and lost environmental resources that are

important to human livelihood. Surely, there can be no serious dispute that such things are of grave moral significance and that a powerful moral obligation exists to avoid errors that cause such effects. The historical argument needs nothing more than this by way of a moral basis. Or, to put the matter another way, it is true that an argument for [the precautionary principle] requires substantive moral premises, but these premises need not take the form of controversial propositions of meta-ethics.[52]

While this seems true, it distorts the sense in which many, Steel included, at least appear to think the precautionary principle has genuinely non-anthropocentric applications. When Steel says that the historical argument for the precautionary principle applies to environmental and human health hazards it is implicit that he is claiming we sometimes ought to protect the environment for its own sake. Many of the historical examples he uses concern threats to human health (e.g., radiation, benzene, and polychlorinated biphenyls – PCBs), but others are framed in terms of their direct environmental consequences (e.g., organochlorine pesticides and the Great Lakes, and tributyltin as an antifouling agent on ships).[53] In presenting the historical argument, Steel discusses the implications of threats of harm to human health or the environment, where the latter concerns appear to be genuinely non-anthropocentric. However, the above explanation of the historical argument appeals to anthropocentric environmental concerns; here he is explicit that he bases his case for the precautionary principle on an argument where moral and instrumental reasoning converge.

Steel does hint toward the end of his book that he has not addressed the tensions that arise between the two main values the precautionary principle is framed to protect, namely the environment and human health. He says that his interpretation of the precautionary principle does not address how much weight should be given, for example, to the health and welfare of agricultural animals as compared to wild animals or humans.[54] Unfortunately this is a major weakness in Steel's account since an interpretation of the precautionary principle cannot be unified if it allows incommensurable versions of the principle. In any case, if his account is to unify work on precaution it must apply to both anthropocentric and non-anthropocentric versions of the precautionary principle since both appear in the vast literature on precaution. Avoiding engagement with the differences between anthropocentric and non-anthropocentric versions of the precautionary principle shirks the deeply philosophical debates in environmental ethics about what has moral standing. Yet given the lack of consensus about what has moral standing and why, it is especially important to be very explicit about the moral claims that underlie precautionary claims.

Steel says that the limited function of the precautionary principle is, "to provide a reasonable framework for decision making that is not susceptible to paralysis by scientific uncertainty."[55] Steel is thus motivated in a case like climate change to find a version of the precautionary principle that will help prevent unjustified delays in taking precautionary action, but in so doing he fails to recognize that so-called versions of the precautionary principle can be incommensurable with one another, undermining the notion that there is one precautionary principle of which this is merely a version. The point about which Steel and I have a substantive disagreement is how precaution should be formalized and utilized. He wants to rescue the precautionary principle as it appears in existing environmental legislation and hence builds a historical argument for his view that makes sense of how "the precautionary principle" has been understood and used in both the real world and academic discussions. I look at the mess and confusion that has arisen around the concept of "the precautionary principle" and think the best way forward is to recognize that there is no (and cannot be a) simple referent to this term; hence we should abandon it.

3 A Way Forward – Rethinking Precaution and Precautionary Principles

Many things could be meant by "the precautionary principle." It could be a general goal. It could be a meta-principle. It could be a moral principle. It could be an approach to risk management. This is precisely the problem. "The" precautionary principle has come to be understood as too many different things. As Kerry Whiteside says, "in spite of talk of the precautionary principle, there is no perfect statement of the principle already out there, just waiting to be found."[56] Yet despite the fact that almost everyone, Whiteside included, continues to talk about "the precautionary principle," many authors also recognize that this term has come to symbolize something much broader.

Whiteside, for example, says, "[t]he precautionary principle is a pragmatically evolving, human principle, born of modern societies' reflections on the nature of new risks."[57] He understands the precautionary principle as guiding us where standard assumptions about risk management do not hold by shifting the default position towards action rather than business as usual in the face of uncertain threats of harm. However, he argues that there can be no 20- or 30-word juridical formulation of the term because what is meant by "the precautionary principle" is this shift in the way we approach decision making. Whiteside's ultimate thesis is that "the precautionary *principle* shades off into precautionary *politics.*"[58] This seems right. There is an important role for precautionary thinking to play; we simply have to abandon the idea that there is a single precautionary principle that can or should guide all such policies.

Along these lines Derek Turner and I have noted that "the precautionary principle" has come to serve as something more like a banner signifying a shared commitment to the welfare of the environment and future persons and, in addition, shared reservations about the effectiveness and applicability of economic cost-benefit analysis.[59] At the same time, the many different ways in which the precautionary principle is understood illustrates, as Whiteside suggests, that we need to rethink what the precautionary principle is and what purposes it serves.

Christian Munthe also appears to be implicitly supportive of this view, focusing on what he calls the requirement of precaution in his search for the normative basis of precaution.[60] From the outset Munthe acknowledges that the so-called precautionary principle has come to dominate public discussions of environmental policy despite ongoing philosophical debates about its merit, nature, and formulations. Munthe's project may be understood as an attempt to find and clarify the "ideal of the desirability of precaution,"[61] as well as "the need for a *morality of precaution*,"[62] which he relates to a theory about "*the moral responsibility of imposing risks.*"[63] In this way Munthe also deemphasizes the precautionary principle while focusing instead on a more general notion of precaution. Munthe suggests his moral ideal of precaution sets out requirements that plausible versions of the precautionary principle must meet. More importantly, he says, "[w]e do not need a precautionary *principle*, we need a *policy* that expresses *a proper degree of precaution.*"[64] So while Munthe does at times discuss precaution in a way that suggests he supports that there are versions of the precautionary principle, hence supporting the family tree view discussed above, he also hints at the need for a reframing of precaution in environmental policy.

In light of all this, we need to abandon the notion that "the precautionary principle" has or even could have a coherent definition, that we could mean something coherent and intelligible by this term. Certainly there are *precautionary principles*, but to call one such principle *the* precautionary principle masks the fact that there are many different kinds of principles that are precautionary in nature. There is no single overarching precautionary principle. We should reject the idea that there is a unified family tree that comprises the precautionary principle and instead see precautionary principles as sharing a kind of family resemblance.[65] There is no principle that describes what constitutes a precautionary principle. That is, while there is no way to unify all precautionary principles, what makes them precautionary principles is that they are precautionary.

Something like Steel's meta-precautionary principle may serve as our guide to identifying precautionary principles, but to identify this meta-precautionary principle *as* the precautionary principle would re-invite confusion about just what "the" precautionary principle is. Even Steel is adamant that his meta-precautionary principle is not *the* precautionary principle. The meta-precautionary principle does not on its own describe what constitutes a precautionary principle. We cannot simply appeal to "the precautionary principle" or "the meta-precautionary principle" to justify precautionary action, since the first is vacuous and the second is not an action-guiding principle. Rather, the meta-precautionary principle is but one type of precautionary principle that can help guide us to other action-guiding precautionary principles of more limited scope, which in turn can guide action in particular cases. A key implication of this view is that each precautionary principle will have to be independently articulated and defended. Much of the literature on "the" precautionary principle will in fact be able to be reinterpreted as attempts to do just this. For example, those who have tried to categorize or limit formulations of "the" precautionary principle may be reinterpreted as identifying different kinds of precautionary principles (e.g., moral precautionary principles vs. epistemic precautionary principles) and suggesting how these principles should be formulated. Those who have defended specific versions of "the" precautionary principle, such as Gardiner and McKinnon,[66] may be understood as defending particular precautionary principles.

Steel challenges this view, claiming that mine is an unstable position. He says:

> For what makes all of these different precautionary principles instances of the same general type? On the one hand, if a substantive answer can be provided to this question, then it seems that some unification of [the precautionary principle] is afoot after all. On the other, if no substantive answer can be provided, then "[the precautionary principle]" would be little more than an empty label that can be applied to almost anything one likes.[67]

That they are precautionary may very loosely unify all precautionary principles, but this does not mean that there is *a* precautionary principle or that there is a meaningful way to strongly unify the precautionary principle. I am even willing to accept that Steel's meta-precautionary principle may capture the essence or limits of precautionary principles, but again this does not mean that all precautionary principles are *versions* of "the" precautionary principle. Precautionary principles are too diverse and potentially incommensurable to be *versions* of a single precautionary principle. They are precautionary in nature, but this does not mean they are all different faces of some single principle.

Steel seems to cling to the phrase "the precautionary principle" in part because he offers a historical argument for his interpretation of the term in light of the fact that too often environmental policy uses uncertainty as an excuse for delay in the face of environmental risks. He argues that the precautionary principle "functions as a corrective intended to move policy making on environmental matters toward greater balance."[68] The precautionary principle, as he sees it, it meant to combat systematic bias and excuses for inaction. Likely because of this, Steel seems to assume that any credible account of the precautionary principle will strengthen institutional appeals to the precautionary principle in a way that does not require a reframing of precaution. However, we need not assume that all precautionary policies are appropriately formulated. Steel thinks it is a criticism of my view that it "would render calls to adhere to [the precautionary principle] found in international environmental agreements vacuous."[69] While this is true, I do not think it is problematic. Many appeals to "the precautionary principle" are vacuous because "the precautionary principle" is vacuous. The way forward is going to involve some work. We are going to have to go back and replace vague references to "the precautionary principle" with specific precautionary principles in environmental and other policies so that it is clear what precautionary approach is suggested in any given case.

While the range of precautionary principles is likely to be vast, I am most interested in identifying moral precautionary principles that capture pro tanto moral reasons.[70] As Shelly Kagan describes, "[a] *pro tanto* reason has genuine weight, but nonetheless can be outweighed by other considerations."[71] In some cases, though certainly not all, we *ought* to take a better-safe-than-sorry approach and act in advance of scientific certainty. We may be pro tanto obligated to take precautionary measures against threats of catastrophe, threats to human subjects of medical testing, and/or threats to endangered species, but we may not be morally obligated to take precautionary measures against breaking a nail or every conceivable threat posed by run-of-the-mill thunderstorms. If a moral precautionary principle is to have any substantive, action-guiding normative force, it must pick out a clear set of moral reasons. Further, our reasons for taking a precautionary approach in certain contexts are usually defeasible, that is they can be overridden by other moral concerns, but this does not mean they are not strong moral reasons; they simply are not the only moral reasons governing our actions. Precautionary principles will often be defeasible in that other moral obligations will sometimes take precedence to precaution, which is why they should be understood as pro tanto moral principles. Exploring the defeasibility conditions of a particular precautionary principle will likely be key to articulating and defending it fully. Pro tanto moral precautionary principles will nonetheless capture cases in which we have strong moral reasons to take precautionary measures.

Yet as action-guiding moral principles, these precautionary principles cannot be perfectly prescriptive in every case, as particular uncertain threats of harm often present unique challenges. However, they can narrowly capture classes of pro tanto obligation or moral reason. That is, as principles, pro tanto moral precautionary principles must allow leeway in their application, but they cannot be so open-ended as to be self-contradictory or practically useless. Carefully formulating precautionary principles so that they capture particular pro tanto moral obligations in an action-guiding way will therefore be both difficult and essential. Drawing on existing work both on how to formulate versions of "the" precautionary principle and on work that is critical of "the" precautionary principle can provide lessons for clearly formulating precautionary principles. For example, to be action guiding, precautionary principles should be formulated in ways that avoid the worries about lack of clarity identified above.

There is also much to be learned from those who have interpreted "the" precautionary principle as a decision-making framework or guide to risk analysis. For example, particular precautionary decision-making frameworks may enable decision makers to determine when a particular precautionary principle suggests precautionary measures are morally warranted and, when this is the case, what precautionary measures would appropriately address the threat of harm in question. Since all precautionary principles address *threats* of harm they hence address cases in which there are a range of uncertainties. However, since no two threats of harm are ever exactly the same, what a precautionary principle requires of us in any given situation will depend on a variety of factors. Because of this, most precautionary principles must have some flexibility for accommodating uncertainty and nuance. This is where precautionary decision-making frameworks may be able to help us implement precautionary principles on a case-by-case basis.

Being responsive to uncertainty also usually entails a precautionary approach to scientific inquiry that is engaged with the policy process because the decision-making process must be in dialogue with the science informing it.[72] That is, precautionary decision making requires paying careful attention to what is known and what is not known about a particular threat of harm and the available precautionary measures for addressing it. Decision makers may call on scientists and analysts to provide information, but scientists and analysts themselves must be responsive to this need and to the dynamic process of decision making in contexts of complexity and uncertainty as well. Joel Ticker therefore calls for methods within and across disciplines that delineate: "what is known and the certainty with which it is known; what is not known; what is suspected; the limits of the science; probable outcomes of different policy options; key areas where new information is needed; and recommended mechanisms of obtaining high-priority information."[73] Tickner goes even further in proposing precautionary assessment, which he describes as a framework and set of procedures for implementing the precautionary principle in environmental and health decision making.[74] Tickner says that precautionary assessment is not meant "to replace existing decision-making structures but rather to reorient them to better support preventive, precautionary decisions in the face of uncertain complex risks."[75] Reframing Tickner's work on precautionary assessment such that it avoids vague talk of "the" precautionary principle enables us to see it as a more general method that could be incorporated into precautionary decision-making frameworks.

4 Conclusion

The intuitions that it is sometimes "better to be safe than sorry" and/or that "we should act in advance of scientific certainty" underlie all discussions and interpretations of the precautionary principle, but it is not clear what this principle is or is supposed to be. "The" precautionary principle is an ambiguous and amorphous idea that has no clear and/or consistent meaning. Some implicitly think of the precautionary principle as a family tree where versions or formulations of it have a shared structure and/or aim; some think of it as a way of approaching decision making in the face of uncertainty; some think of it as a combination of these. However, there is no way to strongly unify *the* precautionary principle. Hence we should abandon the idea that there is *a* precautionary principle and instead rethink the role of precaution, precautionary principles, and precautionary decision-making frameworks.

This will both allow for precaution to have a more meaningful and powerful role in public policy and enable us to better understand our precautionary obligations. We might, that is, understand this interpretation as a whole as articulating the nature of what Whiteside calls precautionary politics. If we are to work on precaution, our project should be to identify and defend precautionary principles that can be applied using precautionary decision-making frameworks in precautionary politics. This interpretation admittedly limits the applicability of "the" precautionary principle in public policy. Policies should no longer generically refer to or appeal to "the" precautionary principle but should instead refer or appeal to specific, independently justified precautionary principles or more broadly appeal to a precautionary approach. It is important that we intervene and change the way we understand and talk about precaution so that we can prevent a confused notion of "the precautionary principle" being used to justify any and all precautionary measures against any type of threat. Reframing how we use and talk about precaution, then, has the potential to return meaning and political force to the concept of precaution, individual precautionary principles, and precautionary decision-making frameworks.

Notes

1 Thalos 2009, 2012; Munthe 2011; Shue 2010; to a lesser extent, McKinnon 2012.
2 Powell 2010, 183.
3 Jordan and O'Riordan 1999, 23.
4 Steel 2015, 2013; McKinnon 2012; Sandin 2007; Sunstein 2005; Posner 2005; Harris and Holm 2002; Sandin 2004; Weckert 2012; Adam Carter and Peterson 2014; Petrenko and McArthur 2009.
5 Cranor 2001, 2004.
6 Gardiner 2006; McKinnon 2012; Randall 2011; Moellendorf 2014.
7 Rio Declaration on Environment and Development. 1992.
8 Raffensperger and Tickner 1999, 8.
9 Patterson and McLean 2010.
10 Grady 2010; Hamilton 2009.
11 Hughes 2006, 452.
12 Peterson 2007; Sandin 2007; Harris and Holm 2002.

13 Manson 2002, 263–74. Others have also identified categorizations of and/or formulas for formulating the precautionary principle, but I find Manson's analysis to be particularly helpful for articulating the ways in which the precautionary principle is unclear. See Steel 2015; Ahteensuu and Sandin 2012; Sandin 2007; Sandin 1999; Trouwborst 2006; Hickey and Walker 1995.
14 Manson 2002, 264.
15 Manson 2002, 264.
16 Manson 2002, 265.
17 Trouwborst 2006; Steel 2015. A simpler interpretation is that the precautionary principle has two parts: a trigger (i.e., risk threshold), and a response (i.e., precautionary response) (Bodansky 2004).
18 Manson 2002, 269.
19 Turner and Hartzell 2004.
20 Hickey and Walker 1995, 423–54.
21 Hickey and Walker 1995, 425–26.
22 Hickey and Walker 1995, 426, see also the elaborated list of these criteria on page 441.
23 Hickey and Walker 1995, 449.
24 Sandin 2007, 109.
25 Turner and Hartzell 2004, 459.
26 Romero-Lankao et al. 2014.
27 Hohmann 1994; Bodansky 2004; Vinuales 2010; Pedersen 2014; Wiener et al. 2011; Freestone and Hey 1996; Marchant and Mossman 2004. For a discussion of the role the precautionary principle has played and can play in climate policy and liability law in particular, see Haritz 2011.
28 Fisher 2002, 28.
29 Petrenko and McArthur 2009, 607.
30 Petrenko and McArthur 2009, 2011.
31 Bodansky 1991.
32 Fisher 2002, 8.
33 Ahteensuu and Sandin 2012, 968.
34 Bodansky 2004, 391.
35 Bodansky 2004, 391.
36 Goklany 2001; Resnik 2003; Harremoes et al. 2002; Commission of the European Communities 2000.
37 Goklany 2001.
38 Goklany 2001, 9.
39 Goklany 2001, 11.
40 Resnik 2003.
41 Resnik 2003, 342.
42 Sandin 2007, 105.
43 Steel 2015, 2013.
44 Steel 2015, 43.
45 Steel 2015, 10.
46 Steel 2015, 18.
47 Steel 2015, 9–10.
48 Steel 2015, 17.
49 Because of this, Steel addresses many topics I do not, such as the connection between the precautionary principle and a rejection of the value-free ideal in science.
50 Steel 2015, 221.
51 Steel 2015, 90.
52 Steel 2015, 92–93.
53 Steel 2015, 70–81.

54 Steel 2015, 210–11.
55 Steel 2015, 211.
56 Whiteside 2006, 150.
57 Whiteside 2006, 150.
58 Whiteside 2006, 150.
59 Turner and Hartzell 2004.
60 Munthe 2011.
61 Munthe 2011, 6.
62 Munthe 2011, 9.
63 Munthe 2011, 10.
64 Munthe 2011, 164.
65 Family resemblance concepts, on Wittgenstein's account, overlap in usage though there is no unifying property bringing them all together (Wittgenstein 1958, sections 67–71). See also Lynch 1999; Forster 2013.
66 Gardiner 2006; McKinnon 2012.
67 Steel 2015, 5–6.
68 Steel 2015, 70.
69 Steel 2015, 5–6.
70 I previously described such precautionary principles as *prima facie* moral principles but have come to agree with Shelly Kagan that *pro tanto* is a better description of the type of moral reasons such principles capture (Kagan 1989).
71 Kagan 1989, 17.
72 Brown 2002; Tickner 2003b.
73 Tickner 2003b, 10.
74 Tickner 2003a.
75 Tickner 2003a, 266.

References

Adam Carter, J., and Martin Peterson. 2014. "On the Epistemology of the Precautionary Principle." *Erkenntnis* 80(1): 1–13. doi:10.1007/s10670-014-9609-x.

Ahteensuu, Marko, and Per Sandin. 2012. "The Precautionary Principle." In *Handbook of Risk Theory*, ed. Sabine Roeser, Rafaela Hillerbrand, Per Sandin, and Martin Peterson, 961–978. Dordrecht: Springer Netherlands.

Bodansky, Daniel. 1991. "Scientific Uncertainty and the Precautionary Principle." *Environment* 33(7): 4–5, 43–44.

Bodansky, Daniel. 2004. "Deconstructing the Precautionary Principle." In *Bringing New Law to Ocean Waters*, ed. David D. Caron and Harry N. Scheiber, 381–391. Martinus Nijhoff Publishers.

Brown, Donald A. 2002. "The Precautionary Principle as a Guide to Environmental Impact Analysis: Lessons Learned from Global Warming." In *Precaution, Environmental Science and Preventive Public Policy*, ed. Joel Tickner, 141–155. Washington, DC: Island Press.

Commission of the European Communities. 2000. "Communication from the Commission on the Precautionary Principle."

Cranor, Carl F. 2001. "Learning from the Law to Address Uncertainty in the Precautionary Principle." *Science and Engineering Ethics* 7(3): 313–326. doi:10.1007/s11948-001-0056-0.

Cranor, Carl F. 2004. "Toward Understanding Aspects of the Precautionary Principle." *The Journal of Medicine and Philosophy* 29(3): 259–279. doi:10.1080/036053104 90500491.

Fisher, Elizabeth. 2002. "Developing a 'Common Understanding' of the Precautionary Principle in the European of European and Comparative Law." *Maastricht Journal of European and Comparative Law* 9(1): 7–28.

Forster, Michael. 2013. "Wittgenstein on Family Resemblance Concepts." In *Wittgenstein's Philosophical Investigations: A Critical Guide*, 66–87. Cambridge: Cambridge University Press.

Freestone, David, and Ellen Hey, eds. 1996. *The Precautionary Principle and International Law: The Challenge of Implementation*. London: Kluwer Law International.

Gardiner, Stephen M. 2006. "A Core Precautionary Principle." *Journal of Political Philosophy* 14(1): 33–60. doi:10.1111/j.1467-9760.2006.00237.x.

Goklany, Indur M. 2001. *The Precautionary Principle: A Critical Appraisal of Environmental Risk Assessment*. Washington, DC: Cato Institute.

Grady, Denise. 2010. "In Feast of Data on BPA Plastic, No Final Answer." *The New York Times*, September 6.

Hamilton, Jon. 2009. "Plastic Peril? Is 'Better Safe than Sorry' Reason Enough for Law?" *NPR*, April 15.

Haritz, Miriam. 2011. *An Inconvenient Deliberation: The Precautionary Principle's Contribution to the Uncertainties Surrounding Climate Change Liability*. Kluwer Law International.

Harremoes, Paul, David Gee, Malcolm MacGarvin, Andy Stirling, Jane Keys, Brian Wynne, and Sofia Guedes Vaz. 2002. *The Precautionary Principle in the 20th Century: Late Lessons from Early Warnings*. London: Earthscan.

Harris, J., and S. Holm. 2002. "Extending Human Lifespan and the Precautionary Paradox." *Journal of Medicine and Philosophy* 27(3): 355–368.

Hickey, James E., and Vern R. Walker. 1995. "Refining the Precautionary Principle in International Environmental Law." *Virginia Environmental Law Journal* 14(3): 423–454.

Hohmann, Herald. 1994. *Precautionary Legal Duties and Principles of Modern International Environmental Law: The Precautionary Principle: International Environmental Law Between Exploitation and Protection*. London: Graham & Trotman.

Hughes, Jonathan. 2006. "How Not to Criticize the Precautionary Principle." *Journal of Medicine and Philosophy* 31(5): 447–464.

Jordan, Andrew, and Timothy O'Riordan. 1999. "The Precautionary Principle in Contemporary Environmental Policy and Politics." In *Protecting Public Health and the Environment: Implementing the Precautionary Principle*, ed. Carolyn Raffensperger and Joel Tickner, 15–35. Washington, DC: Island Press.

Kagan, Shelly. 1989. *The Limits of Morality*. Oxford: Oxford University Press.

Lynch, Michael P. 1999. *Truth in Context: An Essay on Pluralism and Objectivity*. Cambridge: MIT Press.

Manson, N.A. 2002. "Formulating the Precautionary Principle." *Environmental Ethics* 24(3): 263–274.

Marchant, Gary E., and Kenneth L. Mossman. 2004. *Arbitrary & Capricious: The Precautionary Principle in the European Union Courts*. Washington, DC: The AEI Press.

McKinnon, Catriona. 2012. *Climate Change and Future Justice: Precaution, Compensation, and Triage*. New York: Routledge.

Moellendorf, Darrel. 2014. *The Moral Challenges of Dangerous Climate Change: Values, Poverty, and Policy*. Cambridge: Cambridge University Press.

Munthe, Christian. 2011. *The Price of Precaution and the Ethics of Risk*. The Netherlands: Springer.

Patterson, Alan, and Craig McLean. 2010. "Misleading and Dangerous: The Use of the Precautionary Principle in Foreign Policy Debates." *Medicine, Conflict and Survival* 26(1): 48–67.

Pedersen, Ole W. 2014. "From Abundance to Indeterminacy: The Precautionary Principle and its Two Camps of Custom." *Transnational Environmental Law* 3(2): 323–339. doi:10.1017/S2047102514000132.

Peterson, M. 2007. "Should the Precautionary Principle Guide Our Actions or Our Beliefs?" *Journal of Medical Ethics* 33(1): 5–10. http://jme.bmj.com/cgi/doi/10.1136/jme.2005.015495.

Petrenko, Anton, and Dan McArthur. 2009. "Between Same-Sex Marriages and the Large Hadron Collider: Making Sense of the Precautionary Principle." *Science and Engineering Ethics* 16(3): 591–610.

Petrenko, Anton, and Dan McArthur. 2011. "High Stakes Gambling with Unknown Outcomes Justifying the Precautionary Principle." *Journal of Social Philosophy* 42(4): 346–362.

Posner, Richard A. 2005. *Catastrophe: Risk and Response*. Oxford: Oxford University Press.

Powell, Russell. 2010. "What's the Harm? An Evolutionary Theoretical Critique of the Precautionary Principle." *Kennedy Institute of Ethics Journal* 20(2): 181–206.

Raffensperger, Carolyn, and Joel Tickner. 1999. "Introduction: To Foresee and to Forestall." In *Protecting Public Health and the Environment: Implementing the Precautionary Principle*, ed. Carolyn Raffensperger and Joel Tickner, 1–12. Washington, DC: Island Press.

Randall, Alan. 2011. *Risk and Precaution*. Cambridge: Cambridge University Press.

Resnik, David B. 2003. "Is the Precautionary Principle Unscientific?" In *Studies in History and Philosophy of Science Part C: Studies in History and Philosophy of Biological and Biomedical Sciences*.

Rio Declaration on Environment and Development. 1992.

Romero-Lankao, P., J.B. Smith, D.J. Davidson, N.S. Diffenbaugh, P.L. Kinney, P. Kirshen, P. Kovacs, and L. Villers-Ruiz. 2014. "North America." In *Climate Change 2014: Impacts, Adaptation, and Vulnerability. Part B: Regional Aspects. Contribution of Working Group II to the Fifth Assessment Report of the Inter-governmental Panel of Climate Change*, ed. V.R. Barros, C.B. Field, D.J. Dokken, M.D. Mastrandrea, K.J. Mach, T.E. Bilir, M. Chatterjee, et al., 1439–1498. Cambridge and New York: Cambridge University Press.

Sandin, Per. 1999. "Dimensions of the Precautionary Principle." *Human and Ecological Risk Assessment: An International Journal* 5(5) (August 10): 889–907. doi:10.1080/10807039991289185.

Sandin, Per. 2004. "The Precautionary Principle and the Concept of Precaution." *Environmental Values* 13(4): 461–475. doi:10.3197/0963271042772613.

Sandin, Per. 2007. "Common-Sense Precaution and Varieties of the Precautionary Principle." In *Risk: Philosophical Perspectives*, ed. Tim Lewens, 99–112. London: Routledge.

Shue, Henry. 2010. "Deadly Delays, Saving Opportunities: Creating a More Dangerous World?" In *Climate Ethics: Essential Readings*, ed. Stephen M. Gardiner, Simon Caney, Dale Jamieson, and Henry Shue, 146–162. Oxford: Oxford University Press.

Steel, Daniel. 2013. "The Precautionary Principle and the Dilemma Objection." *Ethics Policy Environment* 16(3): 321–340.

Steel, Daniel. 2015. *Philosophy and the Precautionary Principle: Science, Evidence and Environmental Policy.* Cambridge: Cambridge University Press.

Sunstein, Cass R. 2005. *Laws of Fear: Beyond the Precautionary Principle.* Cambridge: Cambridge University Press.

Thalos, Mariam. 2009. "There is No Core to Precaution." *Review Journal of Political Philosophy* 7(2): 41–49.

Thalos, Mariam. 2012. "Precaution Has its Reasons." In *The Environment: Philosophy, Science and Ethics (Topics in Contemporary Philosophy)*, ed. William P. Kabasenche, Michael O'Rourke, and Matthew H. Slater, 171–184. Cambridge: MIT Press.

Tickner, Joel A. 2003a. "Precautionary Assessment: For Integrating Science, Uncertainty, and Preventive Public Policy." In *Precaution, Environmental Science and Preventive Public Policy*, ed. Joel A. Tickner, 265–278. Washington, DC: Island Press.

Tickner, Joel A. 2003b. "The Role of Environmental Science in Precautionary Decision Making." In *Precaution, Environmental Science and Preventive Public Policy*, ed. Joel A. Tickner, 3–19. Washington, DC: Island Press.

Trouwborst, Arie. 2006. *Precautionary Rights and the Duties of States.* Brill.

Turner, Derek, and Lauren Hartzell. 2004. "The Lack of Clarity in the Precautionary Principle." *Environmental Values* 13(4): 449–460. doi:10.3197/0963271042772604.

Vinuales, Jorge E. 2010. "Legal Techniques for Dealing with Scientific Uncertainty in Environmental Law." *Vanderbilt Journal of Transnational Law* 43(437): 437–503.

Weckert, John. 2012. "In Defense of the Precautionary Principle [Commentary]." *IEEE Technology and Society Magazine* 31(4): 12–17. doi:10.1109/MTS.2012.2225464.

Whiteside, Kerry H. 2006. *Precautionary Politics: Principle and Practice in Confronting Environmental Risk.* Cambridge: MIT Press.

Wiener, Jonathan B., Michael D. Rogers, James K. Hammitt, and Peter H. Sand, eds. 2011. *The Reality of Precaution: Comparing Risk Regulation in the United States and Europe.* Washington, DC: Resources for the Future Press.

Wittgenstein, Ludwig. 1958. *Philosophical Investigations.* New York: Macmillan.

3 A Precautionary Approach to Threats of Catastrophes

Why should we take a precautionary approach to climate policy? The answer to this question is simultaneously simple and elusive. Climate change threatens to be harmful in all kinds of ways and so it seems obvious that we should take precautionary measures in an effort to minimize the extent of climate damage. However, it is now clear that we cannot simply appeal to "the" precautionary principle for guidance because there is no single such principle. Instead we need to work out why we should take a precautionary approach in this particular case and whether there are any particular precautionary principles that can guide us. This will enable us to make a strong case for precautionary climate policy and avoid claims that the precautionary principle has failed and should hence be abandoned as a possible guide to climate policy.[1]

One of the most obvious and also forceful reasons we should take a precautionary approach to climate policy is that climate change threatens to be catastrophically harmful to both existing and future humans. I argue in this chapter that we have clear, strong pro tanto moral reasons for taking a precautionary approach in the face of threats of catastrophe. Because precaution is such a broad concept, there are likely many ways in which we could take a precautionary approach to climate change. By focusing on why we have strong pro tanto moral reasons for taking a precautionary approach to threats of catastrophe I hence zero in on the guiding role precaution can and should play in our thinking about climate change from a big picture perspective. My aim is to make it crystal clear that we *ought* to be doing much, much more than we are doing to address the diverse threats of harm climate change poses. Accepting that we have strong moral reasons to take a precautionary approach to threats of catastrophes turns out to be very powerful insofar as it provides a foundation from which aggressive climate policy can be framed.

In this chapter I hence lay the groundwork for precautionary climate policy by defending a precautionary approach to threats of catastrophes comprising a Catastrophic Precautionary Principle and Catastrophic Precautionary Decision-Making Framework, with the first two sections addressing these in turn. I argue that the Catastrophic Precautionary

Principle has to be carefully formulated to avoid the many criticisms that have been leveraged against versions of "the" precautionary principle. The Catastrophic Precautionary Decision-Making Framework, on the other hand, can only suggest the key considerations that must be brought to bear in the decision-making process when we face threats of catastrophe. We have to accept that when faced with complex, uncertain threats of catastrophe, decision making can be well structured, but it will also be inherently messy in all kinds of ways. In the third section I consider several other similar views from the literature so as to reinforce my key arguments and clarify the distinctiveness of my view. Finally, I make explicit the reasons why we can confidently claim that climate change threatens to be catastrophically harmful and hence that the Catastrophic Precautionary Principle may serve as a (partial) guide to climate policy.

1 The Catastrophic Precautionary Principle

While we cannot take precautionary measures against every possible threat of harm, there is something to this intuition. There is a morally relevant difference between taking precautionary measures against threats to ecosystems, protecting human research subjects, and trying to avert a climate catastrophe. Here I am concerned with isolating the third kind of case. My starting point is the simple intuition that if we owe our future selves or future people anything it seems plausible that, all things being equal, we have an obligation to take precautionary measures against foreseeable catastrophes. That is, we have strong moral reasons to take precautionary measures against the worst kind of outcomes, namely those that would be catastrophic. This does not mean that this is the only category of threats of harm we ought to take precautionary measures against, but it is one of the categories that is relatively straightforward to isolate and defend.

Furthermore, the fact that we are sometimes uncertain about the likelihood or nature of possible catastrophes does not undermine the intuition that we *ought* to do something to prevent them given the magnitude of the harm that could ensue. It is also morally irrelevant *when* a catastrophically harmful outcome would happen for us to have strong moral reasons to take precautionary measures against such a possibility – the reason being, if what is bad are catastrophically harmful outcomes, it should not matter who is harmfully affected, be it our families, neighbors, physically distant others, or future generations. Of course, when faced with many threats of catastrophe we may have reasons to prioritize addressing those that threaten to manifest sooner, but this does not undermine the point that we have strong reasons to take precautionary measures against all threats of catastrophe in virtue of there being the possibility of catastrophic harm. Of course, this cannot mean that we ought to take all precautionary measures against every threat of catastrophe. Rather, that an outcome could be catastrophic gives us strong though not defeasible reasons, which need to be spelled out, to take appropriate precautionary measures.

The precautionary principle that applies to threats of catastrophe I defend here is the Catastrophic Precautionary Principle. In order to avoid the concerns Derek Turner and I raise about the lack of clarity in the precautionary principle, the Catastrophic Precautionary Principle is formulated so as to be responsive to the issues discussed in Chapter 2.[2] Namely, it requires that specific parties should be held responsible for implementing and paying for precautionary measures to be determined on a case-by-case basis; it is explicitly anthropocentric, therefore avoiding a problematic tension between protecting human health and the environment for its own sake; it specifies what constitutes a catastrophe; it provides guidance for determining what appropriate precautionary measures are on a case-by-case basis; and it allows for a range of precautionary measures to be considered as appropriate so long as these do not create or exacerbate threats of catastrophe while leaving the judgment of how much precaution is warranted in any given case to be determined on a case-by-case basis. Hence the Catastrophic Precautionary Principle is formulated as follows.

Catastrophic Precautionary Principle

Appropriate precautionary measures should be taken against threats of catastrophe where:

- Catastrophes are outcomes in which many millions of people could suffer severely harmful outcomes (defined as severely detrimental to human health, livelihood, or existence).
- A precise probability of a harmful outcome is not needed to warrant taking precautionary measures so long as the mechanism by which its threat would be realized is well understood and the conditions for the function of the mechanism are accumulating.[3]
- Appropriate precautionary measures must not create further threats of catastrophe and must aim to prevent the potential catastrophe in question.
- Imminent threats of catastrophe require immediate precautionary action.
- Threats of catastrophe that involve an imminent threshold or point of no return for effective precautionary action (beyond which precautionary measures are limited or unavailable) also require immediate precautionary action aimed at preventing this threshold from being crossed.
- Non-imminent threats of catastrophe might warrant further study before further precautionary measures are implemented, provided a delay in taking precautionary measures will not prevent such measures from effectively preventing the catastrophic outcome in question.

- The likelihood (or probability) of a catastrophically harmful outcome may affect *what* precautionary measures are taken but not *that* precautionary measures should be taken. That is, a low-probability outcome/event might warrant more minimal precautionary measures than a similar high-probability outcome/event (which might require aggressive mitigation measures).
- Responsibility for implementing and paying for precautionary measures should be assigned to appropriate parties on a case-by-case basis.

The following five points together explain these elements and constitute a defense of the Catastrophic Precautionary Principle as a plausible pro tanto moral principle.[4]

First I must explain what constitutes a catastrophe and why we have pro tanto moral obligations to take precautionary measures against threats of catastrophe. I define catastrophes as outcomes that are extremely widespread and severe in terms of the number of people who would be harmfully affected and the severity of these effects. I consider a threat extremely widespread if many millions of people are harmfully affected and consider a threat severe if its harmful effects would significantly impact human health, livelihood or existence, as this defines severe harm relative to the most basic human interests.[5] This is a moderate position in that it captures truly large-scale events but does not set the bar so high as only to apply in exceptionally extreme cases. This limits the scope of what we mean by "catastrophe" while ensuring it applies to some of the worst kinds of threats of harm in terms of scope.[6] This definition of catastrophe is admittedly much weaker than those offered by others,[7] but still sets the bar very high.

To see this, imagine a future in which sea level has risen by ten or more meters over the course of a century. Such a dramatic rise would have extremely widespread harmful effects: hundreds of millions of people would be displaced, entire cities and even states would be flooded, and all states with a coastline would suffer significant infrastructure damage. The global social and economic implications, though less direct, would inflict equally dramatic suffering upon many millions of people. Such an outcome would very clearly be catastrophic in the harmful effects it would have, but even less widespread and dramatic harmful outcomes should be considered catastrophic. Widespread droughts in Africa leading to famine causing millions of people to suffer or die, for example, would also be catastrophic in the sense I am using here.

Many irreversible threats of harm, those that are caused by irreversible changes or processes, should also be understood as threats of catastrophe because of the potential unpredictability and permanence of such changes. Irreversible threats of harm need not be immediately catastrophic to generate an obligation to take precautionary measures because of the fact that these harmful effects have the potential to be catastrophic in the long term. If, for example, some natural resource is irreversibly lost, this will be harmful not only to those who depend on this resource now, but also to future people who will never have a chance to benefit from this resource. For example, once a glacier has melted there is no getting it back on human timescales. So any threats of harm posed by melting glaciers such as increased sea level or decreases in fresh water availability have to be considered from an exceptionally long timescale. We therefore should be particularly careful when assessing whether irreversible changes or processes have the potential to be catastrophically harmful in the long run.

The main claim that needs defending, however, is not what constitutes a catastrophe but why we have strong pro tanto moral reasons to take precautionary measures against threats of catastrophe. Why do such threats warrant special moral attention? First and foremost, it is the nature and severity of threats of catastrophe that warrant a precautionary response. It would be wrong to stand by and allow a catastrophe to ensue if we have both sound reasons to believe such an outcome is possible and are in a position to take precautionary action because of the magnitude of the harmful outcome that might otherwise ensue.[8] All we need to agree on to ground the Catastrophic Precautionary Principle is a shared view that catastrophically harmful outcomes are especially bad so that we have strong pro tanto moral reasons to take precautionary action to avoid such outcomes.

If we do not at least have pro tanto reasons to think catastrophically harmful outcomes are bad, and hence that we ought to take appropriate measures to mitigate the risk of such outcomes, I cannot imagine when we would have pro tanto reasons to mitigate risks. Catastrophically harmful outcomes on the definition provided above are simply so bad as to beg for special moral consideration.[9] This entails that we ought to try to anticipate and prevent catastrophes because it is unacceptable to risk many millions of people being severely harmfully affected. Key to this line of reasoning, however, is that such considerations are in the form of pro tanto moral reasons. The pro tanto nature of the Catastrophic Precautionary Principle means that there will be times when morality dictates that we prioritize satisfying other obligations. Preventing catastrophe is not something we always ought to do, as preventing distant or remote catastrophes is unlikely our only or even highest moral priority. If you have to choose between buying lifesaving medicine for your child and donating $100 toward preventing climate catastrophe, I suspect you ought to save your child. However, if a country has to choose between investing in a purely luxury good or spending $100 million on precautionary measures against a climate catastrophe, I suspect it ought to take the precautionary measures and forego the luxury goods for its citizens. The tougher case, of course, will be when a country has to balance providing health care to its poor, educating children, and taking precautionary measures against threats of catastrophe, but even here the Catastrophic Precautionary Principle gives us at least some resources for assessing how strong our precautionary reasons are in a particular context. It is in this sense that we have pro tanto moral reasons to take precautionary measures against catastrophes, though these reasons are defeasible.

This means that it is not just that we have strong reasons to avoid contributing to threats of catastrophe, though this is true by implication, but also that we ought to take precautionary measures against threats of catastrophe regardless of what is causing them. We have the same kinds of strong moral reasons to take precautionary measures against truly natural disasters that threaten to be catastrophically harmful (e.g., tsunamis) as we do to take precautionary measures against anthropogenically caused threats of catastrophic harm (e.g., bioweapons). Climate change, as we have seen, is extremely causally complex in that its effects are not straightforwardly natural or anthropogenic.[10] The causal forces contributing to a threat of catastrophe will undoubtedly prove to be relevant as we sort out what kinds of precautionary measures we ought to take and who should be held responsible for implementing and paying for these measures, but a focus on catastrophic outcomes simplifies the basic justification for taking precautionary measures against threats of catastrophe since it does not require understanding of what causally contributes to a harmful outcome to give us pro tanto moral reasons to take precautionary measures.

While the notion of pro tanto moral reasons may seem abstract, this concept can help us understand the strength and limitations of the Catastrophic Precautionary Principle. A pro tanto moral reason is one that is defeasible by other moral reasons in certain contexts. To say I have a pro tanto moral reason to do something is to say that in general I ought to do it. In most cases pro tanto obligations create strong moral reasons to act in a particular way, but if circumstances dictate that such acts are impossible or conflict with promoting some other (contextually more important) moral obligations, they may be defeated. The upshot of understanding the Catastrophic Precautionary Principle as a pro tanto moral principle is that we can recognize that it provides strong reasons to take precautionary measures against catastrophes without committing us to doing so in all circumstances. The downside of understanding the Catastrophic Precautionary Principle as a pro tanto moral principle is that our reasons for taking precautionary measures are defeasible, and the task of determining when taking precautionary measures should be our moral priority remains to be met. Much of the moral disagreement surrounding the Catastrophic Precautionary Principle will be about its defeasibility conditions, about when we ought to prioritize taking precautionary action against threats of catastrophe in the face of our many other moral demands. For now it is enough to understand that the Catastrophic Precautionary Principle is a pro tanto moral principle that should be widely accepted as such. How strong our precautionary obligation is to try to prevent catastrophes will have to be worked out on a case-by-case basis. As we will see, how likely or imminent a threat of catastrophe is, for example, can affect a reasonable course of precautionary action in any given case.

Second, the Catastrophic Precautionary Principle cannot contain a stringent knowledge condition specifying how much knowledge about a catastrophic threat of harm is needed to generate pro tanto moral reasons to take precautionary measures because this would undermine its applicability to uncertain threats of harm. One of the motivating forces behind any precautionary principle is that sometimes we must act in advance of complete knowledge. Nonetheless we must require some knowledge of a threat before requiring precautionary measures to avoid a version of the precautionary paradox. Henry Shue provides useful guidance here. He argues that when a risk (or threat of harm) has the following three features, we should "ignore entirely questions of probability beyond a certain minimal level of likelihood."[11] These three features are:

1 *massive loss*: the magnitude of the possible losses is massive;
2 *threshold likelihood*: the likelihood of the losses is significant, even if no precise probability can be specified, because (a) the mechanism by which the losses would occur is well understood, and (b) the conditions for the functioning of the mechanism are accumulating; and
3 *non-excessive costs*: the costs of prevention are not excessive (a) in light of the magnitude of possible losses and (b) even considering the other important demands on resources.[12]

Shue's argument is that the seriousness of the magnitude of potential losses justifies taking action to bring the probability that the losses will accrue to as close to zero as possible, assuming this can be achieved without inordinate cost. While Shue himself does not frame his discussion in terms of precaution, his discussion of these first two points provides important insight and grounding to the Catastrophic Precautionary Principle.

Shue is right to argue that when we face a potentially massive loss, we ought, all things being equal, to try to minimize the risk of such loss, even if we do not know the precise probability of the initial risk. The magnitude of the potential loss, here the magnitude of a threat of catastrophe, justifies taking precautionary measures against the threat of harm. However, Shue adds an important caveat here that prevents this view from falling victim to worries about precautionary paralysis. Namely, he limits our precautionary obligations to address threats of massive loss to those cases in which there is a threshold likelihood that the massive loss could actually obtain.[13] Following Shue I therefore formulate the Catastrophic Precautionary Principle so that all that is needed to warrant precautionary measures against a threat of catastrophe is that the mechanism by which the threat would be realized is well understood and the conditions for the function of the mechanism are accumulating. This limits the scope of the Catastrophic Precautionary Principle to avoid worries about its applicability to completely unsubstantiated threats of catastrophe without setting the bar too high. One might worry, for example, about the threats of catastrophic harm posed by an alien invasion, but since we have no credible evidence whatsoever that there are aliens capable of traveling to Earth, nor do we have any idea what an alien invasion would look like, the Catastrophic Precautionary Principle is not triggered here. Shue's criterion ensures that only credible threats of catastrophe will be addressed.

Nonetheless, the Catastrophic Precautionary Principle does not contain anything resembling the third feature, since considering what reasonable costs are is best left out of a pro tanto moral principle. Costs may sometimes be determined to be excessive given other costly moral demands and the severity of burdens precautionary action would put on those determined to be responsible for such action, but considering these costs is not directly part of the pro tanto moral reasoning identified by the Catastrophic Precautionary Principle. Rather, such considerations come into play in the implementation of this principle as it is weighed against other moral obligations. Shue is right to identify excessive cost as a limit on our precautionary obligations to address threats of massive loss, but this concern need not be captured within the pro tanto moral principle that applies to such threats since it will by implication be implicit to the application of such a principle. Given that Shue's project is to address the strong moral reasons we have for addressing climate change now at a general level, it is appropriate that he included the third feature in his discussion of when we ought to act in advance of complete understanding of a risk, but he should have raised the issue of excessive costs separately. However, given that my project is to formalize this general view and articulate and defend the Catastrophic Precautionary Principle as a pro tanto moral principle, I defer addressing economic and other practical considerations to the implementation of this principle.

Third, the Catastrophic Precautionary Principle says very little about what specifically it requires of us, but this is because what constitutes appropriate precautionary measures must be sorted out case by case so that the unique nature of the threat and all relevant information about it can be taken into account. The nature of a threat, how much we know about it, and how likely or imminent it is are all relevant to determining what kind of precautionary measures are appropriate to take. Sometimes a harmful outcome about which we have only limited credible information will at first warrant only minimal research into the nature and likelihood of the threat so as to better understand what precautionary measures are available and which of these might eliminate the threat of catastrophe; at other times a well-understood threat of catastrophe might warrant immediate and aggressive precautionary measures. This is in part why it is helpful to pair the Catastrophic Precautionary Principle with the Catastrophic Precautionary Decision-Making Framework, which is discussed below. This move also responds to worries that "the" precautionary principle always requires the most drastic course of action (e.g., as justifying an invasion of Iraq because of threats posed by weapons of mass destruction).

What the Catastrophic Precautionary Principle requires will depend on many factors. It is *that* it requires us to take some form of precautionary measures that is unwavering. It is formulated so as to provide some guidance as to what appropriate precautionary measures are while explicitly recognizing each threat of catastrophe will require a tailored precautionary approach.

What precautionary measures the Catastrophic Precautionary Principle requires depends on the nature of the threat in question. In particular, imminent threats of harm require immediate precautionary measures whereas non-imminent threats might allow for delayed precautionary measures or a ramping up of precautionary measures. This is to account for the fact that not all threats warrant immediate, aggressive precautionary action. It might be better first to gather information about a temporally distant threat of catastrophe and wait to take aggressive precautionary measures until the threat is better understood or closer at hand, but imminent threats require immediate mitigating action because of their very imminence – once a threat of catastrophe is upon us we must act to prevent catastrophe. This stipulation helps with the worry that the Catastrophic Precautionary Principle is too demanding in some cases by allowing for action to be delayed in non-imminent cases.

Further, threats associated with imminent thresholds for effective precautionary action require immediate precautionary measures, whereas threats not associated with thresholds for effective action might allow for delayed precautionary measures. This is because beyond thresholds, prevention measures are taken off the table; all that remain are precautionary measures that would mitigate the adverse effects of whatever process threatens the harmful outcomes. Consider again the collapse of the West Antarctic Ice Sheet, which will raise sea levels by an estimated 5–6 meters.[14] It is hard to argue that a rise of such magnitude does not threaten to be catastrophic. Until recently, what might trigger or slow down the process of such a collapse, however, was not well understood. In 2007 the IPCC reported that, "no quantitative information is available from the current generation of ice sheet models as to the likelihood or timing of such an event."[15] However, the IPCC was confident that there was some yet unknown temperature threshold after which the collapse would become inevitable: "it appears that mean summer temperatures over the major West Antarctic ice shelves are about as likely as not to pass the melting point if global warming exceeds 5°C, and disintegration might be initiated earlier by surface melting."[16] In 2013 the IPCC reported that the West Antarctic Ice Sheet was shrinking, though there was no evidence that its collapse was at that point inevitable.[17] Unfortunately new evidence suggests otherwise.[18]

The nature of West Antarctica is that once melting has proceeded to a certain point there is nothing that can stop it, not even a reduction in global mean surface temperature. It appears we have already crossed the threshold that will lead to the complete disintegration of the West

Antarctic Ice Sheet, though there are still considerable uncertainties about what all of the contributing factors were to the crossing of this threshold, the timing of melting, and what the consequences of this melting will be. Unfortunately we did not take the possibility of the West Antarctic Ice Sheet's collapse seriously enough, likely because it seemed like such a distant threat. This is why thresholds for effective precautionary action are as important to recognize as threats of catastrophes themselves.

Another example is a flu pandemic. Once a flu virus spreads widely within a population, the range of precautionary measures that can be taken to protect the individuals in this population are limited – prevention is no longer an option. Failing to take precautionary measures before the threshold is crossed amounts to failing to fulfill the obligation to take precautionary measures against the threat.

It is a tragedy and a failing of our precautionary duties that the West Antarctic Ice Sheet appears to be inevitably melting. We knew enough to know that at some point climate change would cross a threshold after which we would not be able to avoid this potentially catastrophic impact. We are left having to hope that adaptive strategies will be able to mitigate at least a great deal of the harmful effects this rise would otherwise cause; the Catastrophic Precautionary Principle certainly demands that we ought to take aggressive precautionary measures against sea level rise. This unfortunate surprise will also, it is hoped, push us to take other thresholds for effective precautionary action more seriously. Especially in the case of climate change, the lesson learned should not be that it is too late to mitigate, but rather that we had better act quickly before we cross additional thresholds beyond which mitigation will be ineffective.

It is also, it is hoped, self-evident why appropriate precautionary measures must not create or exacerbate existing threats of catastrophes, for to do so would be to violate the Catastrophic Precautionary Principle itself. To substitute one catastrophically risky activity for another or to attempt to mitigate or minimize a particular threat of catastrophe by engaging in another catastrophically harmful activity would be to ignore the very pro tanto moral reasons the Catastrophic Precautionary Principle captures. It might be tempting to say that the risks created by a so-called precautionary measure are less than those it is said to combat, but so long as such risks are catastrophic, they are morally unacceptable as a means of acting on the pro tanto moral reasons to combat the primary threat of catastrophe in question. Yet one might worry, as Neil Manson has in a similar context, that this criterion might make the Catastrophic Precautionary Principle self-defeating.[19] Manson's concern is that most precautionary measures themselves could pose catastrophic risks, but since mere possibilities are easy to construct, the Catastrophic Precautionary Principle can require contradictory courses of action. As Manson says, "[w]e could be doomed if we do and doomed if we don't."[20] Here again we see why it is important to limit the scope of the Catastrophic Precautionary Principle so that it only applies to threats of catastrophe where there is sound knowledge of the mechanism by which the catastrophe would be realized and evidence that this mechanism may have been triggered. Precautionary measures that pose or exacerbate other threats of catastrophe are ruled out.

Fourth, there might appear to be a further issue of determining how likely or probable a threat of harm must be in order for it to warrant precautionary action, but this issue is best left to the stage in which we are determining appropriate precautionary action because it is the severity not the likelihood of catastrophic outcomes that grounds our pro tanto reasons for taking precautionary measures. That is, likelihood should not factor into determining whether or not precautionary measures ought to be taken; rather, it should factor into determining what these measures ought to consist in once such an obligation is determined to exist. There are two reasons for this. First, while the sheer magnitude of threats of catastrophe ought to be enough to get our attention and warrant precautionary action, it may be morally relevant both how likely it is that a catastrophic outcome will ensue and the likely timing of such an outcome in terms of the kinds of precautionary measures we should take, especially in light of other morally relevant concerns. For example, we might initially take more aggressive precautionary measures against a very likely threat than a similar, though less likely threat, prioritizing further study of the latter threat before deciding on an appropriate course of precautionary action. This highlights the importance of recognizing research as one possible element of an appropriate course of precautionary action so long as such research is genuinely aimed at informing a more comprehensive precautionary approach. Second, requiring that a threat of harm meet some standard of probability or likelihood would require that we know enough about it to know how likely it is, but again, the Catastrophic Precautionary Principle applies to cases in which there is a large amount of uncertainty. The severity of threats of catastrophe warrants taking seriously all threats of catastrophe that meet the criteria described above (i.e., the causal mechanism is known and the conditions necessary for triggering this mechanism are accruing).[21]

This does suggest that the Catastrophic Precautionary Principle requires us to take precautionary measures against all threats of catastrophe that meet its rather minimal standard of evidence. For example, if we have evidence that an asteroid might be on a collision course with Earth we will have pro tanto reasons to take precautionary measures since we have some historic basis, in addition to other scientific support, for believing such a collision could have catastrophic consequences. One might worry that this makes the Catastrophic Precautionary Principle too demanding, but the stipulation that non-imminent threats of catastrophe do not necessarily require immediately implementing aggressive precautionary measures helps balance worries about low-probability, high-consequence outcomes. Of course, if a non-imminent threat involves a threshold for effective precautionary action, this does not hold. If we have only minimal reasons to think we are in danger of being hit by an asteroid large enough to have catastrophic harmful effects and we have no evidence that such an outcome is imminent, we might determine that the only precautionary action that this threat warrants is the continued funding of astronomical study of whether there are in fact any such asteroids on a collision course with Earth. Only if we determine that an asteroid collision is imminent and/or we identify a threshold for effective precautionary action are we required to take precautionary measures to eliminate the threat of catastrophe (assuming the catastrophe in question can in fact be avoided!).

Fifth, it is worth emphasizing the fact that the Catastrophic Precautionary Principle is an outcome-driven as opposed to an action-driven principle. It tells us that we ought to take precautionary measures against threats of harm because we ought to avoid very bad outcomes. This is because when we think about future threats of harm we should not necessarily think only about outcomes that could result from harmful activities; we should also think about harmful outcomes that could result from natural processes, such as the consequences of a large volcanic eruption. If the same kinds of harmful outcomes were predicted from naturally occurring climatic changes that we predict will actually occur because of anthropogenic climatic changes, we would then have many of the same pro tanto moral reasons to avoid the non-anthropogenic harmful outcomes as we would for avoiding the anthropogenically induced, but otherwise identical, harmful outcomes. The stress here is on outcomes, regardless of whether these outcomes will arise from the actions of humans. Threats of catastrophe are bad and ought to be avoided, period.

It is natural, nonetheless, to want to focus on responsibility for contributing to a threat of harm, and while causal contribution is certainly relevant in determining who ought to bear responsibility for taking precautionary measures against a threat of harm, there is no important difference between anthropogenic and non-anthropogenic threats of harm with respect to whether or not we should take precautionary measures in the first place insofar as both involve threats of catastrophe. That a threat of harm is owing to the intentional actions of humans might add to the reasons for thinking that a potentially harmful outcome ought to be avoided and should be reflected in how responsibility for taking and paying for precautionary action is assigned, but it does not change the account of why precautionary action should be taken in the first place.

In sum, the Catastrophic Precautionary Principle is driven by the magnitude of threats of catastrophe, but it is spelled out in a way that is sensitive to worries about our precautionary obligations being too vague or overwhelmingly demanding. If we take the Catastrophic Precautionary Principle seriously, we will have to accept that we ought to be doing a lot more than we currently are to address the many threats of catastrophe we face every day. We are also going to have to engage in the very difficult task of weighing our precautionary obligations against other moral demands.

2 The Catastrophic Precautionary Decision-Making Framework

The Catastrophic Precautionary Principle sets clear limits on the nature of our precautionary obligations against threats of catastrophe, but it also leaves much to be decided on a case-by-case basis. Precautionary decision making calls for us to think about decision making as an inclusive, multi-disciplinary process to help us deal with the complexity and uncertainty involved in the kinds of cases that beg for a precautionary approach. It does not require a radical shift in decision-making procedures so much as calling for us to codify the ways in which such procedures must be responsive to complex and often uncertain threats of harm in the particular contexts of taking a precautionary approach or implementing a precautionary principle. One of the primary ways in which precautionary decision making can help us accomplish this is by requiring a multidisciplinary perspective that is sensitive to quantitative, qualitative, and normative considerations. The particular precautionary decision-making framework I articulate here is a very specific account of how the paradigm shift to multidisciplinary decision making should be accomplished in the particular context of applications (or potential applications) of the Catastrophic Precautionary Principle.

The Catastrophic Precautionary Decision-Making Framework is meant, first, to provide guidance as to how to determine whether a threat of harm meets the Catastrophic Precautionary Principle's criteria and, second, to determine what precautionary measures should be taken against a particular threat of catastrophe. As I conceive of it, the Catastrophic Precautionary Decision-Making Framework provides guidance on how to assess and address threats of harm. However, sometimes decision makers want to assess whether an activity is potentially harmful, and we can appeal to the Catastrophic Precautionary Decision-Making Framework in such cases as well. This account of the Catastrophic Precautionary Decision-Making Framework is not meant to be required for assessing all threats of harm, as that would lead straight back to the paralysis objection. Rather, it is meant to help decision makers understand whether the Catastrophic Precautionary Principle applies in particular cases or circumstances and what it requires.

The following is an outline of the Catastrophic Precautionary Decision-Making Framework that tries to capture, in an organized and organizing way, the key considerations that should guide the precautionary decision-making process when addressing threats of catastrophe.

The Catastrophic Precautionary Decision-Making Framework

Aims:

- To determine whether the Catastrophic Precautionary Principle supports pro tanto reasons for precautionary action.

- To determine an appropriate course of precautionary action for addressing a threat of harm that has been deemed to require (or warrant) precautionary measures.

Key considerations for achieving these aims:

- Multiple sources of information: draw on all available sources of information.
- Uncertainty: identify and assess uncertainties in knowledge.
- Catastrophe: determine whether there is good reason to think the threat of harm meets the Catastrophic Precautionary Principle's criteria (i.e., is a threat of catastrophe).
- Causation and potential outcomes: identify what is causing and/or contributing to the threat of harm and what the potentially harmful outcomes are.
- Likelihood: assess the likelihood of potential harmful outcomes under different scenarios.
- Available precautionary measures: identify all available courses of precautionary action and the extent to which these measures could eliminate or reduce the threat.
- Thresholds: assess whether there are any points beyond which available precautionary actions are limited.
- Appropriate precautionary measures: determine which of the available precautionary measures are appropriate to address the threat of harm so as to eliminate (if possible) or mitigate the threat.
- Responsibility: identify specific actors who should be held responsible for taking the prescribed precautionary measures.
- Other normative considerations: identify whether there are any other relevant normative considerations that should be taken into account when implementing the proposed plan of precautionary action; modify the proposed plan as necessary.

The Catastrophic Precautionary Decision-Making Framework is by design responsive to all of the issues with which the Catastrophic Precautionary Principle pushes us to wrestle. Yet this description may in many ways seem unsatisfyingly abstract. The challenge is that this framework is meant to apply to a wide range of situations, the nuances of which must be dealt with on a case-by-case basis. One of the most challenging aspects to accept about this decision-making framework, and what makes it so difficult to articulate clearly, is the fact that it applies to cases in which there are often a wide range of uncertainties. It therefore cannot be as precise as, say, cost-benefit analysis or any quantitative, analytical decision-making framework. Whereas in these kinds of decision-making frameworks one can discern what one ought to do by applying a clear formula (e.g., analyze the costs and benefits and, if the benefits exceed the costs, then act), this

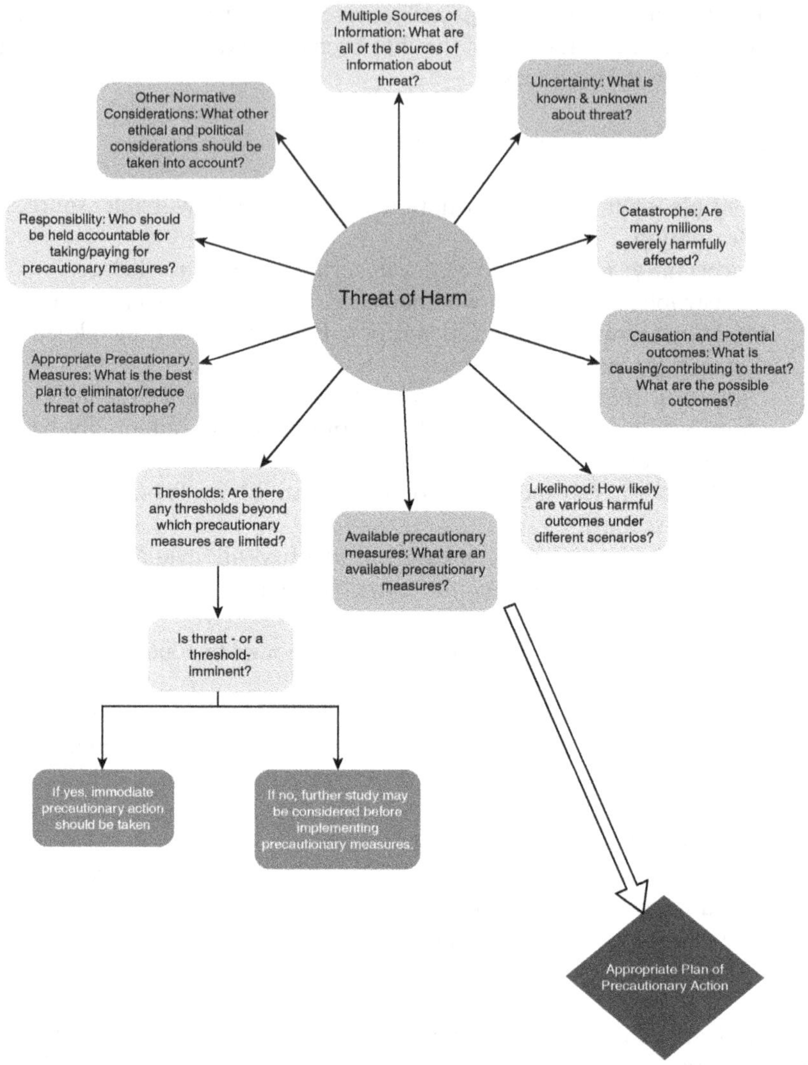

Figure 3.1 Catastrophic Precautionary Decision-Making Framework

Catastrophic Precautionary Decision-Making Framework can only guide us to the hard questions and issues that must be addressed so that judgment calls can be made and appropriate courses of action be decided upon. The upshot of this framework is that it will help contain what is bound to be – and ought to be – a complex, messy decision-making process. As Søren Holm and John Harris say, "[m]any moral choices are complex, and in making political decisions we should not lose sight of this complexity."[22] The Catastrophic Precautionary Decision-Making Framework helps ensure key issues and questions are addressed so that decision makers can avoid the temptation of focusing too narrowly on one set of issues or on available quantitative data when the Catastrophic Precautionary Principle directs us to take seriously uncertain threats of catastrophe.

In order to make the nature of and need for this framework as clear as possible, I discuss each of its ten key considerations in turn. First, in the process of trying to understand the extent and nature of a threat of harm, decision makers should draw on all available sources of information because threats of catastrophe are complex and will often require a multi- and inter-disciplinary approach. Failing to look at all available sources of information about a threat of harm – and considering the possibility that more information should be sought – may lead to either a misunderstanding of the threat and/or a failure to recognize all possible precautionary measures that could be taken against the threat. Precautionary decision making is inclusive in this respect; it also requires taking a broad perspective so as to better understand all aspects of a threat of harm.

Hence the Catastrophic Precautionary Decision-Making Framework supports the kind of approach to scientific research suggested by Joel Tickner's precautionary assessment (discussed in Chapter 2).[23] In order for decision makers to be able to implement the Catastrophic Precautionary Decision-Making Framework scientists and analysts must provide them with the necessary kinds of information and must engage in the policy-making process to know what information is needed and to provide guidance as to what to do with the information they provide.[24] In other words, science cannot be completely divorced from the policy process because the decision-making process must be in dialogue with the science informing it. We must be careful not to place too much emphasis on the role of data or knowledge, however, because the Catastrophic Precautionary Decision-Making Framework is meant to guide decision making even when we lack key information. The more we understand a threat, the easier it will be to decide how we ought to address it, but lack of data or understanding is not a reason for inaction within the Catastrophic Precautionary Principle.

Second, just as it is important to understand what we do know about a threat of harm, it is also important to carefully assess what we do not know about it. This is because just as decision makers should avoid too narrowly focusing on particular bits of knowledge, they must assess the extent to which a threat of harm is understood and what uncertainties remain in our understanding of it. Decision makers must understand both the different kinds of uncertainties involved in our understanding of a threat of harm and how these uncertainties affect our understanding of it in the first place.[25] An analysis of these uncertainties may include assessing, for example, whether there are enough data (and how reliable these data are) to understand fully the harmful outcome in question, how much scientific consensus there is about the nature and/or likelihood of a threat of harm, how confident scientists are in their predictions, whether remaining uncertainty about a threat of harm stems from theoretical challenges (e.g., unpredictability inherent in the system) and/or practical challenges (e.g., lack of data), how other processes affect the likelihood of a threat of harm, and so on. Understanding the role and importance of uncertainty will also vary widely on a case-by-case basis and over time. For example, an assessment of uncertainty about climate change in the 1970s would have been very different from one done today, as a lot more is known about the climatic system and anthropogenic influences on it than 40+ years ago. This suggests that so long as a threat of catastrophe remains, we should continually reassess our knowledge – that is, we should recognize that the Catastrophic Precautionary Decision-Making Framework is not static, as it implicitly recommends an ongoing reassessment of our understanding of both particular threats of catastrophe and what precautionary responses are available to us. Given this, looking to different robustness approaches to decision making and formal uncertainty analyses, which are designed to be responsive to and provide guidance in the face of even genuine (a.k.a., deep) uncertainty, may be helpful, though variations among such approaches should also be taken into account.[26]

Third, upon examining all available information about a threat of harm, as well as what uncertainties remain, decision makers should determine whether there is any reason to think a threat constitutes a threat of catastrophe, whereby many millions of people could be severely harmfully impacted. The reason for this step is simply that in such cases the Catastrophic Precautionary Principle requires precautionary action. As was noted above, particular attention should be paid to irreversible threats of harm because of the potential for these to be catastrophic in the long run.

Fourth, an understanding of what is causally contributing to a threat of harm (or will causally contribute to it in the future) will often help decision makers to better understand the potentially harmful outcome(s) in question and to assess what precautionary measures should be taken. The causal history of a threat of harm can tell us not only what precautionary measures would be effective but also may inform how we think about who should be held responsible for taking and paying for these measures.

Fifth, as I argued above, how likely a threat of harm is does not affect whether we should take precautionary measures against it, but the likelihood of a threat can affect what we deem to be appropriate precautionary measures for addressing it. While the likelihood of uncertain threats of harm in many cases will be unknown (or unknowable) because of uncertainties in our knowledge base, there will nevertheless often be enough information to afford decision makers some sense of its likelihood. With respect to threats of harm that pose different outcomes given different scenarios, decision makers will have enough information, it is hoped, to approximate the likelihood (subjective probability) of each outcome in these different scenarios. How likely a threat of harm appears to be can be relevant when determining what precautionary measures should be taken against it. We should think of our level of understanding and the apparent likelihood of a threat of harm not as information to be used for determining *whether* we are obliged to take precautionary action but as information to be used for determining an appropriate *course* of precautionary action. It is reasonable that we treat the likely threats of harm posed by climate change differently from how we treat the less likely threat of an as-yet unidentified asteroid hitting the Earth.

Sixth, after having assessed the nature and likelihood of a threat of harm, decision makers should identify what precautionary actions are available to address it because the Catastrophic Precautionary Principle does not assume the most extreme course of action is always appropriate. According to this account, anything from ceasing some potentially risky activity to merely gathering more information to determine how to eliminate a threat effectively can count as a precautionary measure. Once again, the Catastrophic Precautionary Principle encourages gathering as much information as possible. Attention should be paid to ensuring available precautionary measures do not violate the Catastrophic Precautionary Principle by creating new or exacerbating existing threats of catastrophe.

Seventh, as argued above, it is particularly important for decision makers to assess whether a threat is imminent and/or there are any thresholds or "points of no return" beyond which a precautionary action can no longer be effective. Beyond these thresholds, prevention measures are taken off the table; all that remain are precautionary measures that would mitigate the adverse effects of whatever process threatens the harmful outcomes. Despite the uncertainties involved, decision makers must carefully assess not only what precautionary measures are available, but also whether there are limits to the timeframes in which such measures would be effective. If it is unknown whether there are points of no return, precautionary measures might include researching whether or not this is the case. Taking preventative precautionary measures against imminent threats and/or before a threshold is crossed will become extremely important if the only way of preventing a harmful outcome will be by staying below the threshold. In such a case, failing to take precautionary measures immediately and/or before the threshold is crossed amounts to failing to fulfill the obligation to take precautionary measures against the threat. However, even when some options for minimizing the harmfulness of a threat remain if action is delayed and/or after a threshold is crossed, decision makers should carefully assess whether such options will be sufficient to satisfy the Catastrophic Precautionary Principle.

Eighth, the most challenging aspect of applying this decision-making framework is determining exactly what precautionary measures should be taken to address a threat of harm – that is, what course of precautionary action is appropriate. This task is challenging because it involves subjectively weighing all of the information gathered about the threat and available precautionary measures against it. Minimally, precautionary measures must aim to bring the threat below the Catastrophic Precautionary

Principle's criteria; that is, they must reduce its harmfulness such that it no longer may be classified as catastrophic. However, just as lesser threats of harm can warrant taking precautionary action, decision makers may decide to set a higher standard.

At first glance it might appear that I am suggesting that decision makers perform a cost-benefit analysis to determine an appropriate course of action. This is not at all what I mean to imply. It is true that decision makers will have to consider the seriousness of the threat of harm and its likelihood and light of available precautionary measures, but their determination must be based on a judgment of what measures would satisfy the moral reasons in favor of precaution. Cost effectiveness or other fiscal considerations can come into play when deciding between equally effective precautionary measures or when other moral considerations are brought in, but such considerations cannot trump the strong pro tanto moral reasons for taking effective precautionary measures in the first place. The hard task of determining what precautionary measures are appropriate – but not excessive – must be left for decision makers to sort out on a case-by-case basis.

Ninth, once appropriate precautionary measures have been identified, responsibility for taking these measures must be assigned to particular actors so as to ensure such measures are actually implemented and that they are implemented in a fair way. There are many different ways in which responsibility for taking precautionary action might be determined: proportionally to causal responsibility, proportionally to ability to pay, according to an equity model, or in any number of other ways. For responsibility too will depend on those features unique to a particular threat of harm, including who or what is causally contributing to the threat. Determining who should be held responsible for taking precautionary action will no doubt be challenging and spark disagreement. The Catastrophic Precautionary Principle does not provide clear guidance on this point because the unique features of a threat of harm will affect how responsibility ought to be understood. This task will involve judging how responsibility for taking precautionary measures should be assigned, which will often need to be informed by other normative considerations such as whether other moral reasons may defeat the pro tanto precautionary reasons for taking precautionary action. I will argue later on, for example, that whenever possible adaptation measures should be implemented in ways that simultaneously satisfy our obligations to assist the development of the global poor. Here again we see that the Catastrophic Precautionary Principle is but one pro tanto moral principle that ought to inform decision making about threats of catastrophe. I see it as a strength of this principle and my approach that I do not attempt to build into this principle a thoroughgoing moral view. Granted that this opens the door to yet more complexity and debate in the decision-making process, but there are so many potential threats of catastrophe, each with their own causal backgrounds, that questions of responsibility for implementing and paying for precautionary measures will have to be context sensitive and bring in other relevant moral considerations.

This brings us to the last consideration, which is that before implementing any plan of precautionary action decision makers should evaluate whether there are other moral considerations that should be accounted for in the implementation of precautionary measures.[27] After all, an appropriate course of precautionary action cannot be decided upon in a vacuum. What other considerations are important will vary from context to context, but decision makers must attempt to take a wide perspective of the normative landscape when judging what precautionary measures ought to be taken because the Catastrophic Precautionary Principle is but one pro tanto principle, which implies that other moral considerations and principles ought to be brought to bear in the decision-making process. As is likely apparent, these issues may trump all of the other issues raised in the precautionary decision-making process. Yet before the implications of the Catastrophic Precautionary Principle can be contextualized in the broader moral landscape, we must understand the nature of a threat of harm and what precaution would recommend in a given case. Only then can we step back and work out whether our pro tanto reason to take precautionary measures stands in light of other moral considerations and/or if we ought to reevaluate what an appropriate course of precautionary action consists in given other moral considerations.

For example, although it is not inherent to the Catastrophic Precautionary Principle itself, whenever possible justice should factor prominently into any implementation of precautionary measures. By all means, precautionary measures should be implemented in a just way. Not only should responsibility for taking precautionary measures be determined by what a fair distribution of this burden may be, but if these measures can be implemented in a way that further promotes the ideals of justice, by all means decision makers should implement precautionary measures in a way that is co-beneficial. The Catastrophic Precautionary Decision-Making Framework should be understood as guiding us to fulfill our obligations to take precautionary measures against threats of catastrophe in ways that simultaneously satisfy as many moral obligations as possible.[28]

That is, the Catastrophic Precautionary Principle should not be considered in isolation. It is important to recognize that we have multiple moral obligations, which will sometimes conflict. For example, the obligation to protect the intrinsic value of nature, if we have such an obligation, will sometimes conflict with the obligation to protect human health. This conflict does not necessarily undermine the status of these obligations; it merely demonstrates that reconciling our many moral obligations will be challenging. This fact, that it is sometimes difficult to reconcile our many moral obligations, is not new to moral philosophy. It reflects on the complexity of our world and in meeting all of our moral obligations simultaneously. We live in a non-ideal world in which millions of people are not having even their basic needs met. This is wrong and we ought to do something about it.[29] However, this does not undermine the fact that we also ought to take precautionary measures against threats of future catastrophe.

The Catastrophic Precautionary Principle raises important questions about how we should balance our obligations here and now against those to future people. Ideally, we will be able to implement our obligations to take precautionary measures against catastrophic threats of harm in a way that coheres with other obligations. Strategies for addressing climate change may have co-benefits for the present generation if such strategies include, for example, helping developing nations to develop and better meet the needs of their people. Given that we have limited resources, however, this will not always be possible. The Catastrophic Precautionary Decision-Making Framework pushes us to recognize the relationships between the Catastrophic Precautionary Principle and other moral obligations, but it alone cannot tell us how to reconcile all of these obligations when they conflict. I leave it open to future research to make progress on this challenge of balancing our obligations to future generations against those to the present generation.

3 The Distinctiveness of My View

Of course, I am not the only theorist who has pointed out that we should take a precautionary approach to threats of catastrophe. Stephen Gardiner, Catriona McKinnon, Cass Sunstein, Darrel Moellendorf, and Daniel Steel all defend principles that are in some ways similar to the Catastrophic Precautionary Principle.[30] Looking at their views reinforces both the argument that we have strong moral reasons to take a precautionary approach to threats of catastrophe but also the distinctiveness of the approach I defend here.

Gardiner argues for what he calls the Rawlsian Core Precautionary Principle, a version of the precautionary principle (or on my view simply *a* precautionary principle) that could be used in environmental policy which is limited by three key Rawlsian criteria for maximin thinking.[31] The Rawlsian Core Precautionary Principle applies when:

1 "[d]ecision-makers either lack, or have reason to sharply discount, information about the probabilities of the possible outcomes of their actions" (i.e., there is uncertainty about the likely impacts of the available actions); and
2 "[d]ecision-makers care relatively little for potential gains that might be made above the minimum that can be guaranteed by the maximin approach" (i.e., the threat of harm in question is so bad as to make any relative alternative benefits pale in comparison); and
3 "[d]ecision-makers face unacceptable alternatives; the threat of harm involves grave risks" (i.e., the threat of harm in question is so bad as to be unacceptable).[32]

Gardiner identifies this as a core version of the precautionary principle because he argues the Rawlsian criteria pick out cases to which the precautionary principle seems clearly to apply. He sees it as an intermediate formulation of the precautionary principle in that it is both determinative (as opposed to optional) and comprehensive (as opposed to exclusive or restrictive of what count as relevant considerations).

While the Rawlsian Core Precautionary Principle does not pick out threats of catastrophe in the way the Catastrophic Precautionary Principle does, it similarly points us to those cases in which we face unacceptably bad threats of harm, the impacts of which are uncertain. So in this sense Gardiner is right to identify the Rawlsian Core Precautionary Principle as being able to "serve as an anchor for the precautionary approach more generally"[33] insofar as it picks out a context in which a precautionary approach is very clearly warranted. The difference between our views is, first, that I do not think Gardiner goes far enough in specifying the scope and limits of the Rawlsian Core Precautionary Principle; his view inherits the internal incoherence many versions of "the" precautionary principle have insofar as he says it applies to environmental policy but does not specify what it is supposed to protect, namely human health and/or the environment; and second, in arguing that the Rawlsian Core Precautionary Principle can serve as an anchor for the precautionary principle, Gardiner seems to maintain a family tree kind of view of "the" precautionary principle, which I have argued is untenable. Nonetheless, Gardiner's view helped moved work on precaution forward. With further specification the Rawlsian Core Precautionary Principle certainly could be *a* strong precautionary principle if Gardiner abandons the idea that it is a version of "the" precautionary principle.

McKinnon defends another Rawlsian view of precaution as it applies to climate change.[34] Her core argument is that the current generation is required to take precautionary measures against climate catastrophes because climate change threatens the possibility of justice for future generations. As she says, "a failure to take precautions puts future generations at risk of suffering unbearable strains of commitment that would make the pursuit of justice impossible for them."[35] Here her argument suggests aggressive mitigation against climate change so as to avoid the passing of key tipping points that could lead to catastrophic climate change. McKinnon distinguishes between weak and strong versions of the precautionary principle, where the weak version is a permissive principle and the strong version is a categorical principle. She notes that in the policy literature it is far more common to see versions of the weak precautionary principle than the strong precautionary principle. Yet she herself is interested in the strong precautionary principle, which she formulates as follows:

The strong precautionary principle: When evidence or information is insufficient to establish the nature and/or probability of harms caused by an activity, policy makers are *required* to act in order adequately to protect people and other entities from these possible harms.[36]

This version of the precautionary principle is strong because it compels policy makers to take precautionary measures even if these turn out to be unnecessary.

McKinnon's first Rawlsian justification for adopting the strong precautionary principle, at least in the case of climate change, is that "the worst consequences of not taking precautionary action are worse than the worst consequences of taking precautionary action"[37] – the key point being that failing to take precautionary measures against climate change would be inconsistent with treating present and future people as equals. McKinnon's second Rawlsian justification for precaution in the face of climate change, which she admits is stronger, is based on more fundamental requirements of intergenerational justice. She argues that if failing to take precautionary measures against climate change "could impose conditions on future generations that would make the joint pursuit of justice by them impossible, then we – the current generation – cannot decide not to take such precautions."[38] Note that while Rawlsian, neither of these reasons appeals strictly to maximin reasoning.

A key similarity between McKinnon's view and my own is that both her strong precautionary principle and my Catastrophic Precautionary Principle are (what I am describing as) pro tanto principles. Neither principle entails that we are necessarily committed to devoting all or even a significant portion of our resources to taking precautionary measures against climate change. McKinnon suggests it is sensible that the strong precautionary principle be qualified with an "all else being equal" clause for much the same reason as I defend the Catastrophic Precautionary Principle as a pro tanto moral principle that requires us to take into account other moral considerations.

The deepest difference between McKinnon's view and my own stems from the nature of our projects: McKinnon aims to describe why justice requires we address climate change, drawing on Rawls's work for inspiration. My aim is to reinterpret "the" precautionary principle, describing a precautionary principle that applies to climate change. McKinnon's discussion of what she calls "the strong precautionary principle" is the result of applying Rawlsian logic to the case of climate change. It turns out that ensuring future generations are not denied the conditions of justice requires taking a precautionary approach to climate change. However, McKinnon does not make any significant claims about precaution in general. In fact, she frequently qualifies that her arguments about the strong precautionary principle as stated only apply in the particular case of climate change. It is only because of this that her discussion of precaution is as strong as it is. The strong precautionary principle, as described by McKinnon, would otherwise fall victim to many of the criticisms facing "the" precautionary principle. This is because though McKinnon offers a very strong, detailed analysis of why we have strong moral reasons to take a precautionary approach to climate change that appeal to the diverse threats of catastrophe climate change threatens, she does not spell out a more detailed version of what she calls "the strong precautionary principle" that is limited to threats of catastrophe. In fact, McKinnon explicitly backs away from making a commitment about whether we have precautionary obligations in cases of non-anthropogenic threats of catastrophe, whereas it is central to my view that we have as much of a reason to take precautionary measures against similar anthropogenic and non-anthropogenic threats of catastrophe.

Despite the differences in our views, I think McKinnon offers a strong argument as to why we ought to take threats of catastrophe seriously: they threaten to undermine the possibility of future generations obtaining the conditions of justice. In my broader project I simply see no need to tie this to activities that impose such threats on future generations. What we would do about the threat of climate change if it in fact were being caused by malicious aliens who were trying to terraform our planet for their own use might turn out to be different from what we actually ought to do to combat anthropogenic climate change, but I argue our *reasons* for taking precautionary measures should be the same in both cases: catastrophes are bad, in part because they threaten the possibility of justice, and hence ought to be avoided.[39]

Sunstein defends yet another view that is similar to the Catastrophic Precautionary Principle. He rejects the precautionary principle in part over worries about precautionary paralysis but defends an Anti-Catastrophe Principle since it is not paralyzing to take a precautionary approach in this limited context.[40] Hence at first glance my view is very similar to Sunstein's. Digging deeper, however, reveals that we approach threats of catastrophes very differently.

To start with, another of Sunstein's reasons for rejecting the precautionary principle is that, at least in cases involving "hazards of ascertainable probability,"[41] cost-benefit analysis (or cost-benefit balancing, as Sunstein sometimes refers to it) can more effectively show what is at stake,[42] though he does not argue that cost-benefit analysis should in and of itself control regulatory decisions.[43] It hence seems that what Sunstein is suggesting is that we should focus on addressing actual costs and risks rather than on being overly precautionary about everything. He says, "[t]he chief advantage of cost-benefit analysis over the precautionary principle is that it provides a wide rather than a narrow viewscreen."[44] This leads him to "suggest that it would be better to endorse cost-benefit analysis while noting that precautions, especially against possible catastrophes, should play a role in its application."[45] While I agree that decision makers should keep a wide viewscreen when they are collecting information relevant to a situation, information in and of itself cannot guide decision making, it can only inform it. Seen as a quantitative tool, cost-benefit analysis is not an alternative to a normative principle such as the Catastrophic Precautionary Principle. Given the nature of Sunstein's concerns, it is not surprising that he rejects the precautionary principle, yet these considerations lead him too to suggest that we have especially strong reasons to take a precautionary approach to threats of catastrophe. Hence he ultimately defends an Anti-Catastrophe Principle, which incorporates his affinity for cost-benefit analysis while also pushing for decision makers to err on the side of caution when faced with catastrophic risks.

It seems that Sunstein's real message here is that the precautionary principle will paralyze if it is used as a guiding principle for everything, but it is a reasonable guide against catastrophic outcomes in the face of uncertainty. Nonetheless, his Anti-Catastrophe Principle has several disadvantages when compared to the Catastrophic Precautionary Principle because it incorporates elements that push us right back to economic considerations. Sunstein qualifies the Anti-Catastrophe Principle such that precautionary measures should be cost effective and that costs matter as such so that the absolute costs of precautionary measures should not be too high. Yet to limit our obligation to take precautionary measures against threats of catastrophe in this way significantly limits the scope of our moral obligations. This is because sometimes it will be very expensive to take precautionary measures against threats of catastrophe, but this does not mean we should not do so. Sometimes avoiding a catastrophic outcome will be worth a high price tag. It is almost ironic that Sunstein's reasons for proposing the Anti-Catastrophe Principle stem from the fact that in the case of uncertain threats of catastrophe, cost-benefit analysis cannot fully inform decision making yet he builds economic considerations into this principle nonetheless. I maintain that economic considerations should not come into play when we are determining what precaution demands because we need first to understand what morality demands of us before we try to figure out how to balance diverse demands given the realities of economic constraints.

The bigger difference between our views, however, is that Sunstein takes himself to be arguing for a maximin approach to threats of catastrophe. He says, "[i]f regulators are operating under conditions of uncertainty, they might well do best to follow maximin, identifying the worst-case scenarios and choosing the approach that eliminates the worst of these."[46] Sunstein takes himself to be following Gardiner here, but is mistaken. Gardiner does draw on Rawlsian maximin criteria in defining and limiting the scope of the Rawlsian Core Precautionary Principle, but this principle is not itself a maximin principle. As McKinnon points out, it is a mistake to conflate the relevant Rawlsian principles with maximin, since maximin is derived from decision theory rather than moral and political reasoning.[47] Sunstein's Anti-Catastrophe Principle requires precautionary measures be taken against the possibility of catastrophe, which will usually be the worst possible threats of harm, but it does not require maximizing the minimum, which in this case would mean something like maximizing the worst possible outcome in the sense of choosing the least bad worst outcome. The problem is that eliminating a worst-case scenario is different from choosing the least bad worst-case scenario. If the Anti-Catastrophe Principle were a maximin principle it would require identifying the worst of all possible outcomes and eliminating these. It would require us to compare the threats of catastrophe posed by climate change, nuclear war, pandemics, asteroid collisions, and even alien invasion and eliminating the worst of these first. To begin with, the amount of uncertainty in our understanding of each of these threats is so great and varying that it seems to be impossible to compare all possible threats of catastrophe against one another so as to create some kind of ordering of worst-case scenarios. More importantly, neither the Anti-Catastrophe Principle nor any version of "the" precautionary principle fully fits the maximin model because precaution does not lead us to ensure that the worst outcome is as good as possible.

If anything, it seems that Sunstein should be identifying his Anti-Catastrophe Principle as a minimax principle. In decision theory "minimax" means to minimize the maximum possible loss. Minimax and maximin principles are related, but there is an important difference between them. Maximin principles aim to raise the bar at the bottom by maximizing minimum gains, while minimax principles aim to eliminate worst-case scenarios. The precautionary principle might be understood as a minimax principle when it is formulated so as to minimize the possibility of the worst threats of harm. Darrel Moellendorf in fact argues for a minimax version of the precautionary principle, which he suggests justifies aggressive climate mitigation.[48] Moellendorf says that the minimax rule, "holds that between courses of action – all with uncertain negative outcomes – the agent should compare only the highest loss scenarios of the courses and choose the course of action that causes the lowest of the highest loss scenarios to come to pass."[49] While this seems like a more plausible interpretation of the precautionary principle, it too faces similar challenges to the maximin interpretation.[50] More importantly for this discussion, a minimax interpretation of the precautionary principle is significantly different from the Catastrophic Precautionary Principle since it ranks possible outcomes relative to one another rather than categorically requiring precautionary measures be taken against a category of threats.[51]

Greg Bognar criticizes Sunstein's Anti-Catastrophe Principle and Gardiner's Rawlsian Core Precautionary Principle, both of which he interprets as maximin versions of "the" precautionary principle.[52] He argues that the Rawlsian conditions for the application of maximin, which guide the formulation of Gardiner's Rawlsian Core Precautionary Principle, do not establish a uniquely rational rule for taking precautionary measures when these conditions apply. This, he suggests, makes maximin versions of "the" precautionary principle dispensable. Bognar rightly points out that it is peculiar and in fact problematic to build into a formulation of the precautionary principle (or what I would call a precautionary principle) a requirement that it only be implemented in cases of uncertainty, for our reasons to take precautionary measures against a threat of catastrophe, for example, do not dissipate as we gain greater certainty about its likelihood of occurrence. Bognar sees Sunstein as having abandoned this first condition about uncertainty. Instead, Sunstein adds clauses to his Anti-Catastrophe Principle that limit its applicability by imposing economic constraints. In essence these caveats limit the principle's applicability to risks because it provides an escape clause for avoiding taking expensive precautionary measures against remote risks of catastrophe. Precautionary principles, however, should apply both when there is uncertainty and to all risks within the prescribed scope.

While I argued in the previous chapter that Daniel Steel fails to unify "the" precautionary principle, Steel is right to point out that there are important similarities between our views. To recap, Steel argues that there are three themes or core elements of the precautionary principle: the meta-precautionary principle, which states that uncertainty should not be used as a reason for inaction in the face of serious threats to the environment or human health; the tripod of a harm condition, knowledge condition, and recommended precaution, which is the three-part structure versions of the precautionary principle must have; and proportionality, which comprises the principles of consistency and efficiency. Together these elements provide guidance about which version of the precautionary principle should be used in a given instance.[53]

Steel argues that I build into the Catastrophic Precautionary Principle the requirement of what he calls the meta-precautionary principle when I specify that all that is needed to trigger the principle is knowledge of a mechanism by which a threat of catastrophe would be realized and that the conditions for the function of the mechanism are accumulating.[54] I agree with Steel that the precautionary intuition that scientific uncertainty should not be used as a reason against taking precautionary measures is part of what weakly unifies all precautionary principles as such. Whether this means that there is a distinct meta-precautionary principle I leave open, but of course the Catastrophic Precautionary Principle, as a precautionary principle, is precautionary in nature. The Catastrophic Precautionary Principle also has the tripod elements Steel (and others) say versions of the precautionary principle should have, though my formulation is longer and more complex than most simplistic three-part statements of so-called versions of the precautionary principle in order to address worries about lack of clarity.

The more interesting point of comparison between our views, as Steel points out, concerns the extent to which my view meets his criterion of proportionality. It is here that I think Steel's view is most helpful in illuminating aspects of the Catastrophic Precautionary Principle in a new way. To begin with, it is clear that the Catastrophic Precautionary Principle satisfies Steel's consistency principle, since it requires that appropriate precautionary measures not create further threats of catastrophe that meet the principle's criteria. The criteria about imminent threats, thresholds for effective precautionary action, and the reasonability of less aggressive measures in the absence of these are also consistent with the consistency and efficiency principles because these aim to rule out superficial, ineffective precautionary measures. The Catastrophic Precautionary Principle also aligns with Steel's vision of proportionality because it requires that the choice of precautionary measures align with its harm and knowledge conditions, in part because how likely a threat of catastrophe is may influence what precautionary measures are deemed appropriate.

Steel's major criticism of my view, however, is that it is overly strict and I go too far in pushing for consistency. Steel rightly calls me out for previously saying, "a precautionary measure cannot introduce a new threat of catastrophe, *however remote or unlikely that threat is.*"[55] This is why I now state the Catastrophic Precautionary Principle such that precautionary measures must not violate the principle itself so that all of the limiting criteria built into this principle also apply to assessing what precautionary measures should be taken. To be consistent precautionary measures should not introduce threats of catastrophe of any likelihood when the mechanism by which the catastrophe would be realized is understood and the conditions for the function of that mechanism are accumulating. My point was and is that appropriate precautionary measures against threats of catastrophe should not introduce new threats of catastrophe that meet the knowledge requirements built into the Catastrophic Precautionary Principle. As Steel says, and I agree, "[t]he threshold of scientific plausibility in an application of consistency is determined by the knowledge condition in the version of [the precautionary principle] used to justify the precaution."[56] Proportionality requires consistency in this way but also that less harmful precautions should generally be preferred over more harmful precautions.

The core difference between our views comes out when we see that part of Steel's vision of "the" precautionary principle is that his core themes help one discern what version of this principle should be applied in any given case, where proportionality plays a central role in this. As he says, "[a]n application of [the precautionary principle] requires selecting a relevant version of [the precautionary principle]."[57] For example, it appears that Steel arrives at the version of the precautionary principle that he applies to climate mitigation by reflecting on what the core themes suggest in the face of climate-affecting activities. He says, "the knowledge condition should be chosen so as to minimize the chance of the harm condition to the extent that is compatible with consistency."[58] I think the idea is that one looks at an issue like climate change that poses threats of harm and formulates a version of the precautionary principle to fit the problem. The version he applies to climate change is, "[i]f a scientifically plausible mechanism exists whereby an activity can lead to a catastrophe, then that activity should be phased out or significantly restricted."[59] Steel suggests that the Catastrophic Precautionary Principle is similar to this, but his version differs substantively from mine. Most glaringly Steel's version applies only to human activities that pose threats of catastrophe, as opposed to all threats of catastrophe regardless of cause, and therefore recommends phasing out or restricting said activity, whereas the Catastrophic Precautionary Principle allows for a much broader range of possible precautionary measures. Because of these differences, Steel's version only applies to climate mitigation, since the activity in question is presumably greenhouse gas emissions (and land use changes), whereas the Catastrophic Precautionary Principle applies to all aspects of climate change and, as the following chapters will make explicit, rightly requires that we carefully consider all possible courses of precautionary action, including adaptation and geoengineering. While Steel's may be *a* precautionary principle, like Moellendorf's, it is much more limited in scope than the Catastrophic Precautionary Principle.

4 The Threat of Climate Catastrophe

It is by now crystal clear that catastrophic climate change is no longer a distant possibility; potentially catastrophic climatic changes are already underway. As Henry Shue said back in 1993, "[t]oday is already the morning after."[60] To establish the applicability of the Catastrophic Precautionary Principle to climate policy I must make explicit the evidence supporting the claim that climate change threatens to be catastrophically harmful. This section hence takes on the first task in applying the Catastrophic Precautionary Decision-Making Framework to climate change. All we need to know to justify precautionary measures against climate change per the Catastrophic Precautionary Principle is the mechanism by which climate catastrophe would be realized and that the conditions for the function of this mechanism are accumulating. In the most basic terms, the relevant mechanism is an amplified greenhouse effect. The conditions for the function of this mechanism are elevated and increasing atmospheric concentrations of greenhouse gases. As of August 2015 the atmospheric carbon dioxide (CO_2) concentration already exceeded 400 parts per million (ppm),[61] which represents an increase of about 40% compared to pre-industrial levels of about 280 ppm.[62] Even on the IPCC's RCP 2.6, which prescribes CO_2 concentrations are limited to 421 ppm by 2100, global mean surface temperatures are predicted to rise between 0.3°C and 1.7°C for 2080–2100 relative to 1986–2005.[63] Remember that this means there is a non-trivial chance that temperatures could be outside this range, possibly being much higher. This implies that warming already in the pipeline (with even an aggressive mitigation strategy – more on this in Chapter 5) poses a significant risk of significantly increased temperatures.

Revisiting the IPCC's reasons for concerns reveals that even at present levels of warming and on this relatively aggressive mitigation pathway we face diverse – and frightening – threats of catastrophe.[64] Looking at the "burning embers" figure provided in Chapter 1 (Figure 1.1) can help us see that we already face the possibility of a range of threats from climate change and that these risks increase significantly with increased global mean surface temperature. Since the Catastrophic Precautionary Principle is activated by any risk of catastrophe, even the IPCC's conservative analysis suggests we have strong reasons to mitigate further climate change and adapt to potential climatic changes already in the pipeline. The IPCC's assessment of the likelihood of even current temperatures posing threats of catastrophe is significant (greater than 66%). This implies that the pro tanto moral reasons we have to take precautionary measures against climate catastrophe are very strong and will be difficult to defeat (i.e., in the face of other moral demands).

The IPCC's first reason for concern stems from impacts to unique and threatened physical, biological, and human systems.[65] These include loss of coastal habitat due to sea level rise, coral bleaching due to sea temperature increases, loss of wetlands from reduced precipitation, and melting of the Arctic ice and permafrost. There is increasingly high confidence that warming of up to 2°C above 1990–2000 levels would have significant impacts on many unique and vulnerable systems, with increasing levels of adverse impacts and confidence at higher levels of temperature increase. However, the IPCC concludes (with medium confidence) that there is moderate risk to many unique and threatened systems below even recent global temperatures. Examples of human impacts in this category include threats to cultures dependent on sea ice and those living on especially low-lying islands. It is not immediately clear whether many millions of people are already affected, though it is clear that as temperatures increase, so will the numbers of people affected. These threats likely do constitute severe threats, however, unless adaptive measures are implemented, since melting sea ice and sea level rise threaten, for example, access to food and water in unique and threatened places.

The second reason for concern stems from the impacts of extreme events, including increasingly intense and frequent heatwaves, tropical cyclones, and intense precipitation events.[66] Just the physical hazards posed by extreme events alone lead the IPCC to conclude (with high confidence) that even recent temperature increases already pose at least moderate risks. Coming decades will bring even more hazardous extreme events (medium confidence). The harmful outcomes associated with these impacts should be understood as catastrophic because of the number of people who would be negatively affected. Large-scale loss of life is among the clearest of all harmful outcomes that ought to be avoided, but the effects on human health that these impacts would have would also be catastrophic if they occurred on a large enough scale, as is predicted with increased warming.

The third reason for concern addresses the distribution of impacts such as decreased crop yields and extreme water shortage that will disproportionately affect particular societies and social-ecological systems.[67]

Such impacts will occur in most regions of the world, but low-latitude and less-developed areas will be most vulnerable. Already there is evidence that climate change is impacting crop production, which has led the IPCC to conclude (with high confidence) even recent temperature increases pose moderate threats. Here again, these threats increase with rising temperatures. Given the direct connection between food production, water shortages and human welfare, and the extent of possible changes even at current temperatures, the IPCC's analysis suggests many millions of people are already at risk of suffering severe harm from climate change.

The fourth reason for concern looks at aggregate impacts and includes risks that can be aggregated into a single global metric including monetary damages or ecosystems lost, the former of which is relevant to this analysis.[68] While there continues to be a low level of agreement about global economy-wide risks (which is reflected in my discussion of the economics of climate change in Chapter 4), there is medium confidence that most people in the world will be negatively impacted with warming of 1°C above 1990–2000 levels due to a combination of aggregate economic risks and biodiversity loss which will contribute to the loss of ecosystem services, with increasing levels of adverse impacts and confidence at higher levels of temperature increase. While I have argued that the Catastrophic Precautionary Principle is a pro tanto moral principle (and should neither be defined in terms of nor constrained by economic considerations), widespread negative economic impacts will likely be associated with other effects that are more clearly catastrophic. If poor states become even poorer this will threaten the ability of their citizens to meet their basic needs (in some cases to a greater extent than is already occurring). The fact that climate change will have widespread economic impacts therefore may be catastrophic because of the secondary impacts of decreasing economic stability and wealth. When aggregate economic impacts are combined with other aggregate impacts (e.g., to ecosystem services), as they are in AR5, it is clear that many millions of people could be severely threatened by rising temperatures, though these impacts may not yet be catastrophic.

The final reason for concern, stemming from large-scale singular events, poses the most obvious threats of catastrophe.[69] These events are sometimes also called "tipping points" or critical thresholds, and involve "abrupt and drastic changes in physical, ecological, or social systems in response to smooth variations in driving forces."[70] Examples of critical thresholds include the collapse of the Atlantic thermohaline circulation, the disruption of the West African or Indian monsoons, or increased amplitude of the El Niño Southern Oscillation,[71] all of which would be catastrophically harmful and irreversible on human timescales. These are also the impacts around which there is the greatest uncertainty. Given this, the IPCC concludes with medium confidence that the risk of singular events becomes moderate above 1°C because of magnitude and irreversibility of the relevant risks primarily stemming from ice sheet collapses, though the possibilities of accelerated methane emissions and the shutdown of the Atlantic meridional overturning circulation (AMOC) are also considered. As I have noted several times, however, there is now evidence that irreversible melting of the West Antarctic Ice Sheet has already begun.[72] The consequent prospect of 4–5 meters of sea level rise most certainly threatens to be severely harmful to many millions of people. This is the clearest threat of catastrophe of all of the reasons for concern because, for example, of the widespread effects several meters of sea level rise would have (e.g., effects on people and infrastructure in coastal areas, salt-water intrusion of groundwater, etc.).

Unfortunately there is clear evidence that the current conditions – observed warming, let alone warming in the pipeline – already pose threats of catastrophe. Climate change already threatens to affect many millions of people because of its diverse and widespread impacts. John Broome has estimated that climate change will cause tens of millions of deaths just this century,[73] which given the long time horizon of climate change is, as John Nolt says, "only the tip of the iceberg" of the climate casualties.[74] So both as a single phenomenon and in regards to the many potentially catastrophically harmful impacts (e.g., impacts of severe drought in a given region over the course of many, many years) climate change threatens to be severely harmful to many millions of people. The Catastrophic Precautionary Principle therefore tells us that we have a pro tanto moral reason to take aggressive precautionary measures against the diverse threats of catastrophe that climate change threatens from both a big picture perspective and with respect to its especially harmful effects. This implies, as we will see, that it gives us strong moral reasons to take a very aggressive approach to mitigation and also to adaptation, as climate change as a whole and many of its impacts threaten to be catastrophically harmful. The burning embers diagram (Figure 1.1) can help us see clearly that even seemingly minimal increases in global mean surface temperature pose catastrophic risks. When seen next to the graph representing even just the likely range of temperature increases for the most conservative RCP (Figure 1.1) (which does not even visualize possibilities outside the 66% probability range), it is crystal clear that the Catastrophic Precautionary Principle provides very strong reasons to take a very aggressive precautionary approach to averting climate catastrophe.

Given the magnitude of harm that climate change threatens it is hard to imagine that the Catastrophic Precautionary Principle will be easily defeated in this case. Certainly there are other moral demands on our resources; many in the present generation are suffering severe harm. However, the fact that climate change threatens to be severely harmful to many, many millions (and likely billions) of people over the course of many, many generations carries serious weight. Of course we should not devote *all* of our resources to combatting climate change, but we certainly ought to be devoting significantly more resources (and attention) to the looming intergenerational tragedy it poses. It turns out, too, that we will often be able to choose those precautionary strategies that simultaneously promote other ends such as sustainable development or women's rights. The Catastrophic Precautionary Principle hence calls for a very aggressive suite of precautionary measures because of the scope, scale, and imminence of the threats of climate catastrophes. The good news is that taking this kind of precautionary approach to climate change is actually affordable, which helps make our strong pro tanto reasons for taking this approach even less defeasible.[75]

5 Conclusion

The Catastrophic Precautionary Principle captures our pro tanto moral reasons to take precautionary measures against threats of catastrophe, climate change included. This precautionary principle captures one kind of case in which the motivating intuition behind talk of "the" precautionary principle stands: we should take precautionary measures against threats of catastrophe, yet it is formulated to be responsive to the worries about the lack of clarity in "the" precautionary principle. We have to be sensitive to the fact that every threat of catastrophe is unique and complex. The Catastrophic Precautionary Decision-Making Framework guides us in the implementation of this principle in such a way that it both acknowledges the necessity of approaching decision making on a case-by-case basis while enabling us to see the Catastrophic Precautionary Principle as strongly action guiding.

This chapter hence suggests a comprehensive precautionary approach to threats of catastrophe. As such, it has the potential to guide us in a wide range of scenarios. For example, when in the fall of 2014 the Ebola outbreak looked like it had the potential to affect many millions of people,[76] this approach could have helped ground an aggressive approach to precautionary action. Looking forward, it may help us prepare for the possibility of any number of pandemics – of either natural or malicious origin. Other possible applications range from the possibility of catastrophic warfare to massive volcanic eruptions that could have global impacts. Here I have established that climate change threatens to be catastrophically harmful so that in the remaining two chapters I am able to set all of these other possibilities aside and focus on the ways in which the Catastrophic Precautionary Principle and Catastrophic Precautionary Decision-Making Framework may guide climate policy. My hope, however, is that the theoretical contributions of this chapter may be helpful in a much wider set of contexts.

Notes

1 Shaw 2009.
2 Turner and Hartzell 2004.
3 This draws directly on Shue 2010.
4 In previous work I identified the Catastrophic Precautionary Principle as a *prima facie* moral principle. I have since come to agree with Kagan that *pro*

tanto is a more accurate description for this type of principle (Kagan 1989), as per the discussion in Chapter 2.

5 I focus on outcomes that would be physically harmful to humans, not on outcomes that would, for example, be harmful to cultural values (e.g., the destruction of a burial ground). There are other kinds of harmful outcomes, but the reasons we have for wanting to avoid such outcomes are different from those for avoiding the physically harmful kinds of outcomes I address here.

6 Admittedly, it will be very difficult to determine when a threat of harm should be deemed catastrophic because concepts such as "health," "livelihood," "extremely widespread," and "severe" are relative and might change over time.

7 For example, Posner 2005. Martin Bunzl's brief discussion of the precautionary principle also addresses the risk of catastrophe where this is understood as involving human extinction. Bunzl only mentions "the most extreme form of the (do nothing unless you can be sure you do no harm) precautionary principle," which suggests a very limited view of "the" precautionary principle that is quite at odds with the view I defend here (Bunzl 2015, 36).

8 Focusing on harmful outcomes rather than on harming per se is important so that we do not get entangled in the complexity of the causal forces at work or the intergenerational nature of climate change (as discussed in Chapter 1 and the Appendix).

9 As is likely apparent, I am not committing to a particular moral theory. I believe that all, or at least most, individualistic, anthropocentric moral theories (and maybe also all moral theories that attribute intrinsic/inherent moral worth to individual humans) will support this view, albeit for different reasons. While this may be unsatisfying to some, I stand by this decision. I myself am a moral pluralist and think that there are many ways to justify the Catastrophic Precautionary Principle, as my discussion in this section suggests. One of my intentions with this book is to make a clear, accessible case for taking precautionary measures against climate change. As such, my project has a pragmatic aim in addition to its philosophical and intellectual aims. I welcome attempts to further ground the views I present here in moral theory.

10 In the language defended in the Appendix this means focusing on catastrophes as a particularly worrisome kind of harmful outcome, therefore avoids entanglement with the concern that our intuitions about *de re* harms do not apply to *de dicto* harms, for we do not at least initially need to understand whether a threat of catastrophically harmful outcome is owing to *de re* or *de dicto* harming to know we have moral reasons to take precautionary measures against it.

11 Shue 2010, 147–48.

12 Shue 2010, 148.

13 See also Gardiner 2006, 2004.

14 Meehl et al. 2007, 819.

15 Meehl et al. 2007, 819.

16 Meehl et al. 2007, 830.

17 Vaughan et al. 2013.

18 Joughin et al. 2014; Rignot et al. 2014; Sumner 2014.

19 Manson 2002.

20 Manson 2002, 273.

21 Together these points help the Catastrophic Precautionary Principle avoid entanglement with concerns about the *de minimis* principle on which sufficiently improbable risks should be ignored. See Adam Carter and Peterson 2014.

22 Holm and Harris 1999, 398.

23 Tickner 2002.

24 Brown 2002; Tickner 2002. See also Weaver et al. 2013.
25 For a perspective on why we need to approach risk management of climate change in ways that can accommodate uncertainty, see Kunreuther et al. 2013.
26 Hall et al. 2012; Lempert et al. 2006; Weaver et al. 2013; Kunreuther et al. 2013; Webster et al. 2012.
27 Whiteside notes an example in actual environmental policy of normative considerations entering into the implementation of "the precautionary principle": in 2000 the European Union added criteria of non-discrimination and coherence to promote fairness in the implementation of "the precautionary principle," in what they call, "Recourse to the Precautionary Principle" (Whiteside 2006, 81–82). The non-discrimination criterion requires that precautionary measures be applied without regard to the geographical location of the harmful effects or of the origin of such effects. The coherence criterion requires that precautionary measures similar in type and scope be taken against comparable threats of harm, even if one is less certain than another. Whiteside argues that transparency and publicity should be added to this list of fairness criteria.
28 For a helpful discussion of the difference between "methods of isolation" and "methods of integration," see Caney 2015.
29 For the argument that the common key to both is alternative energy (eliminating fossil fuels in order to both limit climate change and allow sustainable development), see Henry Shue, "Climate Hope," in Shue 2014.
30 Sunstein 2005; Gardiner 2006; McKinnon 2012; Steel 2015; Moellendorf 2014.
31 Gardiner 2006, 47; see also Rawls 1999, 133–34.
32 Quotes from Gardiner 2006, 47.
33 Gardiner 2006, 34.
34 McKinnon 2012.
35 McKinnon 2012, 50.
36 McKinnon 2012, 54.
37 McKinnon 2012, 56.
38 McKinnon 2012, 63.
39 Another difference between McKinnon's and my own view is that she argues we should contribute to an intergenerational climate change compensation fund that will both pay for both adaptation and compensate those who are in fact harmed by climate change, whereas I will argue in Chapter 5 that adaptation and compensation should not be so lumped together.
40 Sunstein 2005.
41 Sunstein 2005, 59.
42 For a discussion of four key weaknesses of Sunstein's rejection of the precautionary principle in favor of cost-benefit analysis, although ultimately also critical of the precautionary principle, see Clarke 2010.
43 Sunstein 2005, 130.
44 Sunstein 2005, 130.
45 Sunstein 2005, 130.
46 Sunstein 2005, 109.
47 McKinnon 2012.
48 Moellendorf 2014.
49 Moellendorf 2014, 81.
50 For a discussion of why the precautionary principle should not be understood as either a maximin or minimax principle, see also Steel 2015.
51 Moellendorf provides a strong argument that minimax precautionary thinking may be particularly apt in the case of climate mitigation, but he does not provide a strong argument for a minimax precautionary principle in general. His version of the precautionary principle is so different from the Catastrophic Precautionary Principle that it does not compete with it, though it may be

another precautionary principle that can be independently justified and applied in particular circumstances.

52 Bognar 2011. Bognar's solution for salvaging the precautionary approach is to suggest the plausibility of a prioritarian approach that is similar to maximin versions of the precautionary principle, as he understand them, but it offers more flexible reasons for giving priority to avoiding the worst outcomes. I neither find Bognar's suggestion to be in competition with my view of the Catastrophic Precautionary Principle nor particularly promising in its own right (for many of the reasons stated in this chapter), so I do not offer a thorough account of it here.

53 Steel 2015.

54 Steel 2015, 50.

55 Hartzell-Nichols 2012, as quoted in Steel 2015, 51, emphasis added by Steel.

56 Steel 2015, 52.

57 Steel 2015, 30.

58 Steel 2015, 202.

59 Steel 2015, 28.

60 Shue 1993, sec. 43.

61 NOAA 2015.

62 IPCC 2013, 2007.

63 IPCC 2013.

64 The IPCC's conclusions discussed in this section all appear in Oppenheimer et al. 2014.

65 Oppenheimer et al. 2014, sec. 19.6.3.2.

66 Oppenheimer et al. 2014, sec. 19.6.3.3.

67 Oppenheimer et al. 2014, sec. 19.6.3.4.

68 Oppenheimer et al. 2014, sec. 19.6.3.5.

69 Oppenheimer et al. 2014, sec. 19.6.3.6.

70 Oppenheimer et al. 2014, 1079.

71 See also Lenton 2011.

72 Joughin et al. 2014; Rignot et al. 2014; Sumner 2014.

73 Broome 2012, as cited in Nolt 2014.

74 Nolt 2014.

75 van den Bergh 2009. Note that this article also further defends the kind of precautionary approach to climate policy defended in this book.

76 Centers for Disease Control and Prevention 2014. Note that this report anticipates the possibility of 1.4 million cases of Ebola by January 20, 2015. This implies the possibility of many millions of cases beyond that date given the rapid spread of Ebola in this scenario.

References

Adam Carter, J., and Martin Peterson. 2014. "On the Epistemology of the Precautionary Principle." *Erkenntnis* 80(1): 1–13. doi:10.1007/s10670-014-9609-x.

Bognar, Greg. 2011. "Can the Maximin Principle Serve as a Basis for Climate Change Policy?" *Monist* 94(3): 329–348.

Broome, John. 2012. *Climate Matters: Ethics in a Warming World.* W.W. Norton & Company.

Brown, Donald A. 2002. "The Precautionary Principle as a Guide to Environmental Impact Analysis: Lessons Learned from Global Warming." In *Precaution, Environmental Science and Preventive Public Policy*, ed. Joel Tickner, 141–155. Washington, DC: Island Press.

Bunzl, Martin. 2015. *Uncertainty and the Philosophy of Climate Change.* New York: Earthscan.

Caney, Simon. 2015. "Just Emissions." *Philosophy & Public Affairs* 40(4): 255–300. doi:10.1111/papa.12005.

Centers for Disease Control and Prevention. 2014. "Questions and Answers: Estimating the Future Number of Cases in the Ebola Epidemic – Liberia and Sierra Leone, 2014–2015." Centers for Disease Control and Prevention. www.cdc.gov/vhf/ebola/outbreaks/2014-west-africa/qa-mmwr-estimating-future-cases.html.

Clarke, Steve. 2010. "Cognitive Bias and the Precautionary Principle: What's Wrong with the Core Argument in Sunstein's Laws of Fear and a Way to Fix It." *Journal of Risk Research* 13(2): 163–174. doi:10.1080/13669870903126200.

Gardiner, Stephen M. 2004. "Ethics and Global Climate Change." *Ethics* 114: 555–600.

Gardiner, Stephen M. 2006. "A Core Precautionary Principle." *Journal of Political Philosophy* 14(1): 33–60. doi:10.1111/j.1467-9760.2006.00237.x.

Hall, Jim W., Robert J. Lempert, Klaus Keller, Andrew Hackbarth, Christophe Mijere, and David J. McInerney. 2012. "Robust Climate Policies under Uncertainty: A Comparison of Robust Decision Making and Info-Gap Methods." *Risk Analysis: An Official Publication of the Society for Risk Analysis* 32(10): 1657–1672. doi:10.1111/j.1539-6924.2012.01802.x.

Hartzell-Nichols, Lauren. 2012. "Precaution and Solar Radiation Management." *Ethics, Policy & Environment* 15(2): 158–171. doi:10.1080/21550085.2012.685561.

Holm, S., and J. Harris. 1999. "Precautionary Principle Stifles Discovery." *Nature* 400(6743): 398. doi:10.1038/22626.

IPCC. 2007. "Summary for Policymakers." In *Climate Change 2007: The Physical Science Basis. Contribution of Working Group I to the Fourth Assessment Report of the Intergovernmental Panel on Climate Change*, ed. S. Solomon, D. Qin, M. Manning, Z. Chen, M. Marquis, K.B. Averyt, M. Tignor, and H.L. Miller. Cambridge: Cambridge University Press.

IPCC. 2013. "Summary for Policymakers." In *Climate Change 2013: The Physical Science Basis. Contribution of Working Group I to the Fifth Assessment Report of the Intergovernmental Panel on Climate Change*, ed. P.M. Midgley, T.F. Stocker, D. Qin, G.-K. Plattner, M. Tignor, S.K. Allen, J. Boschung, A. Nauels, Y. Xia, V. Bex, 1–30. Cambridge: Cambridge University Press. doi:10.1017/CBO9781107415324.004.

Joughin, Ian, Benjamin E. Smith, and Brooke Medley. 2014. "Marine Ice Sheet Collapse Potentially Under Way for the Thwaites Glacier Basic." *Science* 344 (6185): 735–738.

Kagan, Shelly. 1989. *The Limits of Morality.* Oxford: Oxford University Press.

Kunreuther, Howard, Geoffrey Heal, Myles Allen, Ottmar Edenhofer, Christopher B. Field, and Gary Yohe. 2013. "Risk Management and Climate Change." *Nature Climate Change* 3(5): 447–450. doi:10.1038/nclimate1740.

Lempert, Robert J., David G. Groves, Steven W. Popper, and Steve C. Bankes. 2006. "A General, Analytic Method for Generating Robust Strategies and Narrative Scenarios." *Management Science* 52(4): 514–528. doi:10.1287/mnsc.1050.0472.

Lenton, Timothy M. 2011. "Beyond 2°C: Redefining Dangerous Climate Change for Physical Systems." *Wiley Interdisciplinary Reviews: Climate Change* 2(3) (May 10): 451–461. doi:10.1002/wcc.107.

Manson, NA. 2002. "Formulating the Precautionary Principle." *Environmental Ethics* 24(3): 263–274.

McKinnon, Catriona. 2012. *Climate Change and Future Justice: Precaution, Compensation, and Triage*. New York: Routledge.

Meehl, G.A., T.F. Stocker, W.D. Collins, P. Friedlingstein, A.T. Gaye, J.M. Gregory, A. Kitoh, et al. 2007. "2007: Global Climate Projections." In *Climate Change 2007: The Physical Science Basis. Contribution of Working Group I to the Fourth Assessment Report of the Intergovernmental Panel on Climate Change*, ed. S. Solomon, D. Qin, M. Manning, Z. Chen, M. Marquis, K.B. Averyt, M. Tignor, and H.L. Miller. Cambridge: Cambridge University Press.

Moellendorf, Darrel. 2014. *The Moral Challenges of Dangerous Climate Change: Values, Poverty, and Policy*. Cambridge: Cambridge University Press.

National Oceanic and Atmospheric Administration (NOAA). 2015. "Global Greenhouse Gas Reference Network." Accessed August 11, 2015.

Nolt, John. 2014. "Casualties as a Moral Measure of Climate Change." *Climatic Change* 130(3) (April 26): 347–358. doi:10.1007/s10584-014-1131-2.

Oppenheimer, M., M. Campos, R. Warren, J. Birkmann, G. Luber, B. O'Neill, and K. Takahashi. 2014. "Emergent Risks and Key Vulnerabilities." In *Climate Change 2014: Impacts, Adaptation, and Vulnerability. Part A: Global and Sectoral Aspects. Contribution of Working Group II to the Fifth Assessment Report of the Intergovernmental Panel of Climate Change*, ed. C.B. Field, V.R. Barros, D.J. Dokken, K.J. Mach, M.D. Mastrandrea, T.E. Bilir, M. Chatterjee, et al., 1039–1099. Cambridge: Cambridge University Press.

Posner, Richard A. 2005. *Catastrophe: Risk and Response*. Oxford: Oxford University Press.

Rawls, John. 1999. *A Theory of Justice*, revised edition. Cambridge: Harvard University Press.

Rignot, E.J., J. Mouginot, M. Morlinghem, H. Seroussi, and B. Scheuchl. 2014. "Widespread, Rapid Grounding Line Retreat of Pine Island, Thwaites, Smith, and Kohler Glaciers, West Antarctica, from 1992 to 2011." *Geophysical Research Letters* 34(2): 3502–3509.

Shaw, Chris. 2009. "The Dangerous Limits of Dangerous Limits: Climate Change and the Precautionary Principle." *Sociological Review* 57 (October): 103–123.

Shue, Henry. 1993. "Subsistence Emissions and Luxury Emissions." *Law & Policy* 15: 39–59.

Shue, Henry. 2010. "Deadly Delays, Saving Opportunities: Creating a More Dangerous World?" In *Climate Ethics: Essential Readings*, ed. Stephen M. Gardiner, Simon Caney, Dale Jamieson, and Henry Shue, 146–162. Oxford: Oxford University Press.

Shue, Henry. 2014. *Climate Justice: Vulnerability and Protection*. Oxford: Oxford University Press.

Steel, Daniel. 2015. *Philosophy and the Precautionary Principle: Science, Evidence and Environmental Policy*. Cambridge: Cambridge University Press.

Sumner, Thomas. 2014. "No Stopping the Collapse of West Antarctic Ice Sheet." *Science* 344(6185): 683.

Sunstein, Cass R. 2005. *Laws of Fear: Beyond the Precautionary Principle*. Cambridge: Cambridge University Press.

Tickner, Joel A. 2002. "Precautionary Assessment: A Framework for Integrating Science, Uncertainty, and Preventive Public Policy." In *Precaution, Environmental Science and Preventive Public Policy*, ed. Joel A. Tickner, 265–278. Washington, DC: Island Press.

Turner, Derek, and Lauren Hartzell. 2004. "The Lack of Clarity in the Precautionary Principle." *Environmental Values* 13(4): 449–460. doi:10.3197/0963271042772604.

van den Bergh, Jeroen C.J.M. 2009. "Safe Climate Policy Is affordable – 12 Reasons." *Climatic Change* 101(3–4): 339–385. doi:10.1007/s10584-009-9719-7.

Vaughan, D.G., J.C. Comiso, I. Allison, J. Carrasco, G. Kaser, R. Kwok, P. Mote, et al. 2013. "2013: Observations: Cryosphere." In *Climate Change 2013: The Physical Science Basis. Contribution of Working Group I to the Fifth Assessment Report of the Intergovernmental Panel on Climate Change*, ed. T.F. Stocker, D. Qin, G.-K. Plattner, M. Tignor, S.K. Allen, J. Boschung, A. Nauels, Y. Xia, V. Bex, and P.M. Midgley. Cambridge: Cambridge University Press.

Weaver, Christopher P., Robert J. Lempert, Casey Brown, John a. Hall, David Revell, and Daniel Sarewitz. 2013. "Improving the Contribution of Climate Model Information to Decision Making: The Value and Demands of Robust Decision Frameworks." *Wiley Interdisciplinary Reviews: Climate Change* 4(1): 39–60. doi:10.1002/wcc.202.

Webster, Mort, Andrei P. Sokolov, John M. Reilly, Chris E. Forest, Sergey Paltsev, Adam Schlosser, Chien Wang, et al. 2012. "Analysis of Climate Policy Targets under Uncertainty." *Climatic Change* 112(3–4): 569–583. doi:10.1007/s10584-011-0260-0.

Whiteside, Kerry H. 2006. *Precautionary Politics: Principle and Practice in Confronting Environmental Risk*. Cambridge: MIT Press.

4 Precaution and the Economics of Climate Change

Much of political decision making today is driven by economics. It is therefore important to address the argument that we do not need separate precautionary principles like the Catastrophic Precautionary Principle because any genuine precautionary consideration can be built into economic analyses. Richard Posner writes:

> The "Precautionary Principle" ("better safe than sorry") popular in Europe and among Greens generally is not a satisfactory alternative to cost-benefit analysis, if only because of its sponginess – if it is an alternative at all. In its more tempered versions, the principle is indistinguishable from cost-benefit analysis with risk aversion assumed. Risk aversion, as we know, entails that extra weight be given to the downside of uncertain prospects. In effect it magnifies certain costs, but it does not thereby overthrow cost-benefit analysis, as some advocates of the Precautionary Principle may believe.[1]

Posner's understanding of the precautionary principle here echoes the sentiments of many economists. The precautionary principle, according to this view, is nothing more than risk aversion, which cost-benefit analysis can incorporate.[2] While it is hopefully clear by now that there is no one precautionary principle, which may be part of the reason why Posner and others find it to be so spongy, it is worth addressing the view that economics supplants the need for a principle like the Catastrophic Precautionary Principle especially as it arises in debates over how to assess the economic impacts of climate change. For if economics could incorporate the kind of precautionary approach captured by the Catastrophic Precautionary Principle it would likely turn out that we do not need to appeal to it or the Catastrophic Precautionary Decision-Making Framework to determine an appropriate precautionary approach to climate policy; we could instead simply turn to economics to guide us. The key problems are that economic analyses cannot adequately incorporate concerns about uncertain threats of harm and will always be driven by economic thinking as opposed to moral values (i.e., focused on costs rather than moral obligation).[3]

The first section of this chapter explains why economic assessments have a hard time assessing climate change because of uncertainty. First, when there are possible outcomes about which we are genuinely uncertain, there is a danger that economic assessments will either ignore or fail to appropriately consider these outcomes, whereas the Catastrophic Precautionary Decision-Making Framework can nonetheless guide decision making. Second, when we know enough to calculate climate risks, yet uncertainties persist (e.g., about the timing or extent of climate damages), economic assessments can provide useful (though not definitive) information and recommendations but should not be the sole guide to decision making. Here the advantage of the Catastrophic Precautionary Decision-Making Framework is that it forces us to assess all information that is available about threats of harm as well as the limits of this information in a way most economic assessments ultimately fail to do.

To illustrate the force of these arguments the second section of this chapter describes how the above plays out in debates about the economics of climate change. I focus on three distinct approaches to the economics of climate change: William Nordhaus's very traditional cost-benefit analysis,[4] Nicholas Stern's multi-dimensional methodology that attempts to move beyond cost-benefit and build in a precautionary approach,[5] and Gernot Wagner and Martin Weitzman's approach that emphasizes the significance of possible "tail disasters."[6] What becomes apparent is that while Nordhaus and Stern at first appear very divergent in their treatment of the economics of climate change, Weitzman is the theorist who offers the starkest departure from the more traditional cost-benefit paradigm that Nordhaus most closely follows. Stern's work appears quite precautionary when compared to that of Nordhaus, but Weitzman's criticism is that both of them are missing a key precautionary consideration to assessing the costs of climate change appropriately. The third section addresses the practice of discounting in the economics of climate change. I argue that uncertainty provides an argument against the practice of discounting in economic assessments of climate change altogether because it can mask morally relevant outcomes in the distant future. Even economists who take a seemingly precautionary approach to uncertainty, such as Stern and Wagner and Weitzman, fail to realize that merely using a low discount rate is not enough because doing so masks consumption streams that may be morally relevant. In the end, however, even Weitzman's approach falls short of what the Catastrophic Precautionary Principle requires of us, revealing key limitations to any economic assessment of climate change.

In sum this chapter serves as a further defense of the precautionary approach to threats of catastrophe established in the previous chapter as it specifically applies to the decision-making processes that should guide climate policy. The next, final chapter explores in greater detail what precautionary climate policy should look like.

1 Uncertainty's Uncertain Effect on Economic Assessments of Climate Change

Uncertainties abound in our understanding of climate change. As we saw in Chapter 1, climate change is a complex phenomenon that functions globally over extremely long time horizons. Yet despite this we know a lot about climate change. Recognizing the many ways in which uncertainty enters into our understanding of climate change is critical if we are to draw appropriately on this understanding to assess how we should address climate change. This applies as much to economists as it does philosophers, and many economists have devoted significant attention to discussing how to assess the economic costs of climate change in the face of diverse uncertainties.

The two forms of uncertainty at work here are what I call genuine uncertainty and the gray area of uncertainty. Genuine uncertainty generally tracks what many economists think of simply as uncertainty. As was discussed in Chapter 1, risk is usually defined in economics as randomness with knowable probabilities, whereas uncertainty is randomness with unknowable probabilities.[7] I use the term *genuine uncertainty* (which is similar to what others have called *deep uncertainty*) as applying both to those cases in which there is randomness with unknowable probabilities as well as cases in which we simply do not yet know probabilities. So outcomes about which we cannot quantify probabilities all fall into the category of genuine uncertainty. The *gray area of uncertainty* falls somewhere in between Knight's neat categories of risk and uncertainty. In the gray area we have enough knowledge of possible outcomes to calculate probabilities (or make likelihood estimates) but limitations in data and methods are such that some uncertainty nonetheless underlies these calculations. The first two subsections here address these forms of uncertainty in turn, and the third addresses their overlap. Section 3 addresses the implications of all this for the practice of discounting in economic assessments of climate change.

1.1 Genuine Uncertainty

The argument that economic assessments cannot address or adequately account for genuinely uncertain threats of harm and therefore should not guide decision making in such cases is quite a simple one to make. Simply put: probabilities for genuinely uncertain outcomes cannot be meaningfully quantified and hence cannot be meaningfully incorporated into quantitative assessments of climate risk.[8] Yet in the case of climate change there are many gaps in our knowledge of where genuine uncertainties lurk. In many cases we have strong scientific reasons to think an outcome is possible but either do not have enough data or do not understand clearly enough all the causal mechanisms involved to be able to reliably calculate how likely a possible outcome is. As Wagner and Weitzman point out, "[t]he underlying models do their best to capture the 'known unknowns,' and even there they miss quite a bit. By definition, they don't yet capture the 'known unknowns.'"[9] Unfortunately in the case of climate change it is really hard to adequately capture the "known unknowns" and impossible to model the "known unknowns" that probably exist. To lean on language from the Catastrophic Precautionary Principle, in many cases we recognize there may be a mechanism by which a certain kind of potentially catastrophically harmful climatic impact may be realized and know that the conditions for the function of this mechanism are being realized (either simply via GHG emissions or through some secondary mechanism such as a feedback loop), yet we do not know enough to quantify the likelihood of such an outcome with any confidence.

To see this we can return to the example of the West Antarctic Ice Sheet presented in the Introduction. Not that long ago this scenario represented an example of a threat of harm about which we had some information – enough to identify the possible outcome as potentially catastrophic, but about which we knew too little to quantify probabilities with any confidence. Even now, while the inevitable melting of the entire West Antarctic Ice Sheet is quite possibly already underway, genuine uncertainty persists as to the timing of melting under different warming scenarios. Yet how quickly melting occurs will likely significantly affect how harmful the eventual sea level rise it contributes to is. Even before recent data came to light suggesting irreversible melting of the West Antarctic Ice Sheet may be underway, however, we had reasons to worry about this possibility. We knew enough to recognize that once started, it very likely would be impossible to stop and we had no meaningful sense of where the threshold for irreversible melting might be or that we might already have crossed it. How can we take into account possible outcomes like the melting of the West Antarctic Ice Sheet if we cannot reliably calculate how likely they are to occur given different (emissions or temperature) scenarios? Even now, but certainly back as recently as 2010, there were no meaningful or reliable probability estimates for the melting of the West Antarctic Ice Sheet or related climate damages. Yet to leave these out of an assessment of the costs of climate change would lead to significant underestimations of the social cost of carbon.

Climate change does not follow the rules of traditional economic assumptions, which ignore the potential for irreversible changes and thresholds for effective action.[10] Yet we should not just leave out of our analyses the possibility of genuinely uncertain threats of harm, especially when these are catastrophic, merely because our analytical framework is ill-equipped to incorporate them. As Mariam Thalos says:

> Concern with irreversibility is thus a species of concern that some of our risk is undertaken with large exposure and little cushioning. This is why special caution is in order. And in risk analysis as it currently stands, there is no distinction between risks we undertake while cushioned and risks we undertake while not cushioned.[11]

The Catastrophic Precautionary Decision-Making Framework requires us to keep such possibilities in mind while assessing all available information because of our pro tanto reasons for taking precautionary measures against potentially catastrophic outcomes. We must ensure, and not just assume, that we *will* be able to compensate for lost resources or harmful damages. This is why we require a careful analysis of the nature of a threat of harm and what precautionary measures will be able to prevent (or minimize the risk of) catastrophically harmful outcomes.

The Catastrophic Precautionary Principle in conjunction with the Catastrophic Precautionary Decision-Making Framework is valuable here in part because it pushes us to see both the importance of genuinely uncertain outcomes (when these represent threats of catastrophic harm) and to understand whether or how such outcomes are being treated in economic assessments. For as David Fleming says in assessing the relationship between the precautionary principle and economics, "[i]gnorance of the extent of a risk should not be mistaken for presumption of no risk; nor should it be mistaken for presumption of no significant risk. These statements must have tenure in any evaluation of environmental policy options, whatever the cost implications may be."[12] While we will see some attempts to address this issue in the examples of economic approaches to climate economics below, it is critical that we do not let genuine uncertainties drop out of the decision-making process. Yet it is hard to see how economic assessments of climate change that are driven by quantitative methods can adequately account for genuinely uncertain outcomes. It is therefore problematic for decision makers to lean too heavily on economic analyses to inform their thinking. While here it may be helpful to look to robust decision-making frameworks and formal uncertainty analyses that are designed to address just such genuine uncertainties, we should not lean too hard on any quantitative or quantitatively driven assessments because of the challenges associated with modeling genuinely uncertain outcomes.[13] Economic assessments of climate change are at risk of masking their treatment of genuinely uncertain possible outcomes or, worse, ignoring unquantifiable outcomes altogether. This would be a mistake from a precautionary perspective.

1.2 The Gray Area of Uncertainty

While genuine uncertainty limits our ability to quantify climate risks, much of our knowledge of possible climate impacts falls into a gray area between (Knightian) risk and uncertainty. Here we have enough information to assign probabilities to different outcomes given different emissions scenarios but there is still some uncertainty inherent to these probability calculations. The IPCC hence uses a system for identifying the likelihood of outcomes and their collective confidence in their results to reflect both statistical uncertainty (e.g., a range of model results) as well as underlying uncertainties in their data or methods (as discussed in Chapter 1). Further-more, economists not only need to look at the likelihood of future tempera-tures for different scenarios but also at what climate damages are expected at different temperatures. Uncertainties multiply here since for a given emissions scenario or RCP there is a likely range of temperature outcomes, each of which has a likely range of climate impacts. Robert S. Pindyck has in fact argued that the most uncertain aspect of climate change from an economic perspective concerns climate damages or economic impacts.[14]

The challenge for economists is that because such predictions fall into what I am calling the gray area of uncertainty they have to work not with precise probability calculations but with a range of possible outcomes given a particular scenario. There is a way in which no matter how economists choose to treat the uncertain data when assessing climate change, their results will in some way inherit the uncertainties of the scientific assessments on which they draw. For again, not only are there uncertainties about climate sensitivity (how much global warming will ensue given a doubling of CO_2) but economists also need to know what climate damages will be associated with a given temperature scenario, which invites further uncer-tainties because of a range of factors such as how quickly temperatures rise (since faster rates of warming will likely be more harmful than slower warming).

From a precautionary perspective it is problematic that economic assessments of climate change can appear misleadingly precise when in fact they fail to fully take into account the complete spectrum of possible outcomes captured by the scientific data and results on which they draw. As we will see below, there are different ways to accommodate the gray area of uncertainty, some of which are more precautionary than others. An advantage of the Catastrophic Precautionary Decision-Making Frame-work is that it pushes us very carefully to recognize and think about how uncertainty plays a key role in our understanding of climate impacts so that we can pay attention to how this uncertainty is treated in economic and other assessments of climate risk.

1.3 Overlapping Uncertainties

Of course, in reality the gray area of uncertainty and genuine uncertainties cannot be cleanly separated from one another. Both types of uncertainty are present in almost all aspects of our understanding of climate change. Considering the following example will help illustrate the interconnections between and implications of these different forms of uncertainty about climate change for economics. Figure 4.1 represents the IPCC's projections of global surface warming over the next century given different emissions scenarios as of 2007 from AR4.[15] For each emissions scenario, the IPCC gives a likely range (gray bars to the right of the graph) for surface warming in 2100 along with a best estimate (solid line within the gray bars) of what this will be. Given the way likelihood statements were defined in AR4, this means that as of 2007 the IPCC concluded that there is a 66% probability that surface warming in 2100 will fall within the ranges identified by the gray bars for the given emissions scenarios. Indeed there is a gray area of uncertainty.

Looking beyond the bars, however, reveals the way in which genuine uncertainties are relevant here too. While it is tempting to focus only on the visually represented, likely, range, there is a 34% chance that surface warming will fall *outside* this range. There is not only genuine uncertainty about what climate damages would look like at very high temperatures but also genuine uncertainties about potential climate feedbacks which could affect how much warming will be caused by each scenario represented above. For example, as ice sheets and glaciers melt, surface albedo changes because water and (especially vegetated) land are much darker than ice. So in addition to all of the genuine uncertainties about ice sheet and glacial melting, there are genuine uncertainties about subsequent albedo

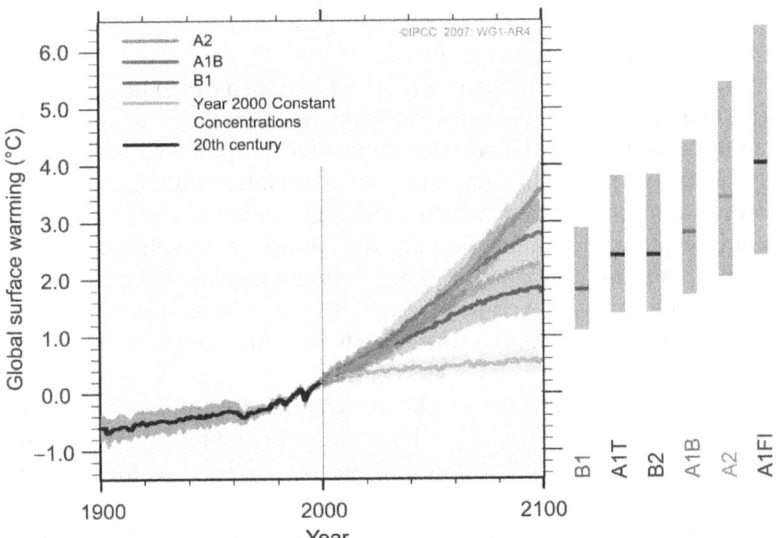

Figure 4.1 Multi-model Averages and Assessed Ranges for Surface Warming
Source: IPCC 2007, 14
Note: Solid lines are multi-model global averages of surface warming (relative to 1980–99) for the scenarios A2, A1B and B1, shown as continuations of the twentieth-century simulations. Shading denotes the ±1 standard deviation range of individual model annual averages. The orange line is for the experiment where concentrations were held constant at year 2000 values. The gray bars on the right indicate the best estimate (solid line within each bar) and the *likely* range assessed for the six Special Report on Emissions Scenarios (SRES) marker scenarios. The assessment of the best estimate and *likely* ranges in the gray bars includes the Atmosphere-Ocean General Circulation Models (AOGCMs) on the left of the figure, as well as results from the hierarchy of independent models and observational constraints.

changes that might increase surface warming. There are also genuine uncertainties about ocean circulation patterns and how or whether these will be affected and in turn further affect global average surface temperature. The complexity of the climate system, as discussed in Chapter 1, raises many questions about possible feedback mechanisms that we do not yet understand well enough to fully incorporate into all of our climate models. Some of these uncertainties inform the predicted likely ranges of temperature outcomes (which partly accounts for why such wide ranges are predicted), while others are left out entirely.

Any economic assessment that draws on data with error bands like the IPCC's 2007 projections of surface warming over the next century given different emissions scenarios (different climate policies) must find a way to use such projections despite the fact that the IPCC identifies ranges of likely temperatures (and only gives a "best estimate" for specific degrees of warming) and the fact that there is a 34% probability, according to the judgment of the IPCC, that the actual warming will be outside these ranges. When we add to this that these predictions likely do not fully or adequately account for all possible but genuinely uncertain feedback, the economists' task is daunting indeed.

What economists ultimately assess are the costs or damages associated with different emissions pathways (that includes both an assessment of climate damages and the costs of achieving the pathway in question). So they need to know what damages are associated with each temperature outcome and then assess which emissions scenario (or policy pathway) should be pursued. Again, though, this choice is complicated by the fact that there is a range of possible temperature scenarios associated with each emissions pathway as well as a range of likely climate impacts associated with each temperature scenario – both of which are limited by genuine uncertainties.

Quantitative assessments of climate change are influenced by the gray area of uncertainty because of limitations in data (e.g., from imperfect temperature records) and methods (e.g., models that average large areas when in fact there are significant local variations), whereas genuine uncertainties about possible climate outcomes are not represented in our models at all. When economists draw on quantitative assessments of climate change they have to make choices about what numbers to use and what to do about the possibilities about which we do not even have good numbers. Economic assessments hence inherit the uncertainties of the data they draw on to drive calculations of climate damages and economically sensible policy pathways. Recognizing this even in the abstract should be enough to raise questions about the extent to which we should lean on economics as a guide to decision making if we take seriously the precautionary obligations to which the Catastrophic Precautionary Principle commits us. The Catastrophic Precautionary Decision-Making Framework can guide us to see exactly this point, which is why climate economics should inform but not on its own guide decision making about climate policy.

2 Uncertainty and the Economics of Climate Change in Practice

What does economics say about climate change? This is like asking what philosophy says about justice. To both questions there are a wide range of answers, many of which deeply conflict with one another. All economic assessments of climate change use quantitative methods to assess climate policy pathways and make recommendations based on what these assessments suggest about their economic consequences. However, there is significant variation in how quantitative methods are used, how the costs and benefits of climate change are quantified, and whether or how additional qualitative concerns enter into assessments. What is most revealing in debates of the economics of climate change for present purposes is *that* there are such divergent approaches and that even those approaches that seem quite precautionary fall short of aligning with the Catastrophic Precautionary Principle. Looking at three core approaches to the economic assessment of climate change reinforces the argument that we should maintain a separate Catastrophic Precautionary Principle and be guided by the Catastrophic Precautionary Decision-Making Framework so that we maintain the ability to distinguish between the different approaches, appeal to those that are most appropriate in a given context, and not lose sight of the moral reasons that should be guiding us in the first place. The first subsection introduces these approaches, and the second discusses their treatment of uncertainty.

2.1 Three Very Different Economic Assessments of Climate Change

A quick overview of William Nordhaus's, Nicholas Stern's, and Gernot Wagner and Martin Weitzman's approaches to the economics of climate change highlights how different economists are coming to very different conclusions and are making very different recommendations for climate policy. In 2008 Nordhaus's "best guess" based on his modeling, despite the large uncertainties involved, was that "the economic damages from climate change with no interventions will be on the order of 2.5 percent of world output per year by the end of the twenty-first century."[16] This relatively low figure led Nordhaus to suggest a conservative approach to climate change policy that involves a gradual "ramping up" to significant emissions reductions. In his 2013 book Nordhaus seems to take a more aggressive approach because of a greater emphasis on thresholds and worst-case options, though he continues to emphasize the importance of balancing the costs of abatement against climate damages in a way that suggests a relatively moderate approach to climate policy.[17]

The *Stern Review*, Stern's 2007 publication, on the other hand, concludes, "if we don't act, the overall costs and risks of climate change will be equivalent to losing at least 5% of global GDP [gross domestic product] each year, now and forever. If a wider range of risks and impacts is taken into account, the estimates of damage could rise to 20% of GDP or more."[18] The *Stern Review* concludes that the cost of acting to prevent the worst impacts of climate change – reducing greenhouse gas emissions – would only constitute around 1% of global GDP each year and hence recommends relatively aggressive climate change policies. In his 2015 book Stern advocates even more strongly for an aggressive approach to climate policy, emphasizing that quantitative assessments "grossly underestimate risk, the social cost of carbon, the necessary urgency in climate action, and the potential attractiveness and discovery of the alternative low-carbon pathways."[19] Stern hence has devoted significant energy to quantitatively assessing climate change while at the same time emphasizing the limits to even his own approach.

Wagner and Weitzman (and Weitzman on his own) go even further in stressing the limitations of quantitative economic assessments of climate change. In the end their message is that economists and all of us should be focusing our attention on the possibility of extreme climate change, on what are sometimes called "tail disasters" or "fat tails" because of the shape of the curve of probability density functions of the likelihood of high warming outcomes in climate models. Wagner and Weitzman are driven by the idea that "[m]ost everything we know tells us climate change is bad. Most everything we don't know tells us it's probably worse."[20] Their ultimate recommendations include working hard to quantify as many possible climate damages as possible while also ensuring that a precautionary stance towards uncertainties about catastrophic risks plays a key role in guiding decision making about climate change.

These different stances are driven by very different approaches to analyzing the economics of climate change. Nordhaus uses the DICE model (Dynamic Integrated model of Climate and the Economy) to evaluate the costs and benefits of climate change. This model draws upon optimal growth theory for its analysis. Different climate change policies "are evaluated on the basis of their contribution to the economic welfare (or, more precisely, consumption) of different generations."[21] This is accomplished by examining how economies make investments in capital, education, technologies and the natural capital of the climate system – where concentrations of greenhouse gas emissions are viewed as negative natural capital and emissions reductions are investments that increase natural capital – in order to increase consumption in the future (at the expense of present consumption).[22] Nordhaus, drawing on a range of disciplines, describes DICE as a model linking, on a global scale, economic growth, CO_2 emissions, the carbon cycle, climate changes, climatic damages, and climate-change policies. DICE assumes that the world has a "well-defined set of preferences" that ranks possible consumption paths, which is represented by a "social welfare function."[23] These consumption paths are determined by two main factors in the DICE model, namely "the overall savings rate for physical capital and the emissions-control rate for greenhouse gases."[24]

Nordhaus argues that his analysis offers important information about climate change, while admitting that it should not be the sole basis of decision making. He says: "In practice, an economic analysis of climate change weighs the costs of slowing climate change against the damages of more rapid climate changes."[25] Nordhaus uses a basic model that is as "simple and transparent" as possible, while incorporating the latest economic and scientific knowledge.[26] He admits that the aim of quantitative economic analyses of climate change is not to provide definite answers, since the inherent uncertainties involved make this impossible, but rather to provide internally consistent answers and "at best provide a state-of-the-art description of the impacts of different forces and policies."[27] As he says, economic models "are not truth machines."[28] Nordhaus is even more explicit in his 2013 book that while there are some limits to quantitative assessments of climate change, his core approach is very much a cost-benefit analysis model.

The *Stern Review*, on the other hand, explicitly identifies the limitations of cost-benefit analysis, taking "a broad view of the economics required to understand the challenges of climate change."[29] It acknowledges that cost-benefit analysis can provide useful information about climate change but also that it has limitations that require using cost-benefit analysis merely as a starting point for further work.[30] The *Stern Review* argues that there are four aspects of climate change that, together, are difficult to accommodate in standard economic theories of externalities and the standard welfare-economic approach to policy. These are: (1) the causes and effects of climate change are global; (2) climate change's impacts are persistent and develop over time; (3) there are considerable uncertainties about the size and timing of impacts as well as about costs of climate change; and (4) impacts will significantly affect global economy, so analysis must consider potentially non-marginal changes to societies.[31] The *Stern Review* acknowledges that modeling the quantitative implications of different climate change policies requires considerable simplification. It also requires making "explicit decisions about the ethical framework appropriate for aggregating costs and benefits of action."[32] This is why the *Stern Review* takes a multi-dimensional approach to the economics of climate change and uses the PAGE (Policy Analysis of the Greenhouse Effect) model, which better accommodates risk and uncertainty than the DICE model. The *Stern Review* argues, however, that a separate precautionary principle need not be imposed as an extra ethical criterion because the *Review*'s analytical approach incorporates aspects of insurance, caution and precaution directly.[33] In other words, the *Stern Review* uses precautionary assumptions such that a separate precautionary principle is unnecessary.

Despite attempting to expand its approach beyond traditional cost-benefit analysis, the *Stern Review* is deeply rooted in economic thinking and identifies climate change as "the greatest example of market failure we have ever seen."[34] In justifying the need for aggressive climate change mitigation the *Stern Review* says, "[t]he benefits of doing more clearly outweigh the costs. Delay would entail more climate change and eventually higher costs of tackling the problem."[35] So in the end the *Stern Review* leans heavily on cost-benefit analysis as its primary methodology, though as we will see, it builds in non-traditional assumptions at various points resulting in very different recommendations from Nordhaus. Stern's later book seems to push even harder on the importance of going beyond cost-benefit analysis, as he offers a stark criticism of integrated assessment models like DICE that combine climate science with economic modeling.[36]

Wagner and Weitzman, however, criticize Stern for very similar reasons to those for which Stern criticizes Nordhaus. They imply that Stern fails to see the limitations of even a much more expansive and inclusive approach to the economic analysis of climate change. Wagner and Weitzman argue that, "[s]ince we know that fat tails can dominate the final outcome, the decision criterion ought to focus on avoiding the possibility of these kinds of catastrophic damages in the first place."[37] This sounds a lot like the Catastrophic Precautionary Principle and in fact Wagner and Weitzman explicitly note that some people call their position a "precautionary principle" (their quotation marks). The key difference between their core precautionary view and my own is in how we are conceptualizing catastrophic damages. When describing fat tails Wagner and Weitzman note: "The IPCC says it's 'very unlikely' that climate sensitivity is above 6°C (11°F). That's comforting but for its definition of just what 'very unlikely' means: a chance of anywhere between 0 and 10 percent."[38] Their point is that a 10% chance of 6°C of warming is not insignificant. In fact, it is possible that tail disasters like this should be a key focus of economists because of how damaging 6°C of warming would be.

It is not clear exactly how Wagner and Weitzman define "catastrophic damages" but their discussion of fat tails focuses on the risk of exceeding 4.5°C, 6°C, or more. While 6°C of warming would be devastatingly catastrophic, climate change threatens to be catastrophically harmful with even much less warming than this. Wagner and Weitzman are worried about scenarios that would be economically catastrophic, not about what catastrophically harmful outcomes we should worry about from a moral perspective. They claim to be very concerned with ethics, but ultimately they are working within an economic mindset that frames the issue in terms of costs, so they conceptualize catastrophes in economic terms (as outcomes that would be very costly), whereas the Catastrophic Precautionary Principle is based on a moral concept of catastrophe that sets the bar much lower. Even Wagner and Weitzman's seemingly precautionary perspective is therefore not nearly as demanding as the Catastrophic Precautionary Principle.

2.2 Addressing Uncertainty (or Not) in Economic Assessments

All of the above economists acknowledge that there are uncertainties about climate change. Even Nordhaus concedes that "[i]n the present context, we have a complex system that is imperfectly understood in the sense that we are unsure how the system will evolve in the future. The uncertainty is based on incomplete knowledge about external variables and about the system itself."[39] The question is how these economists address this uncertainty in their models and interpretation of their results.

Nordhaus concedes that it is very difficult to reliably estimate the economic damages of climate change. Nevertheless, in his 2008 book he talks about how the DICE model addresses this uncertainty because it assumes that although the damages from small and gradual climate change would be modest, damages will rise nonlinearly with increasing global mean surface temperatures.[40] This stems from Nordhaus's treatment of both gray and genuine uncertainties about climate change. Returning to Figure 4.1 (from the IPCC's AR4) provides a starting point for understanding how Nordhaus deals with uncertainty. In his 2008 analysis Nordhaus says that he tries to account for uncertainty as carefully and transparently as possible. In calculating the social cost of carbon, for example, he provides uncertainty bands for his predictions that stem from his use of a range of predicted climate outcomes, like those represented in Figure 4.1 above.[41] In order to draw specific conclusions and make specific recommendations, however, Nordhaus must use, for example, a specific price for the social cost of carbon, which he judges to be very close to the mean in the uncertainty band produced by his analysis. This is akin to looking first at the IPCC's likely temperature range for a given emissions scenario (the gray bars in Figure 4.1) and then focusing on the IPCC's "best guess" (represented by the graphed temperature changes). Looking at the IPCC's likely range is useful but it leaves out the 34% chance that the outcome will not fall within this range. Only considering the IPCC's "best estimate" incorporates the best judgment of the IPCC but ignores the fact that the IPCC does not even assign a specific probability to this best guess. While Nordhaus's analysis is not quite so simple, the point is that while he accounts for uncertainty, his results ultimately must narrow in on specific possibilities and limited ranges for possible outcomes. This does not mean that his results are uninformative, but it does show that they rest on many layers of judgment in addition to objective calculations. Importantly, Nordhaus finds it reasonable to focus on the most likely outcomes rather than worst-case scenarios when running his quantitative models.

Nordhaus himself acknowledges that "[i]t is generally necessary to use judgmental probabilities in analyses of climate change because there are limited or no historical observations on which to base assessments of the parameters of concern."[42] He admits, too, that there is no single methodology for determining these judgmental probabilities. Furthermore, Nordhaus admits that uncertainty enters into his cost-benefit analysis of climate change at almost every stage:

> [T]he structure, equations, data, and parameters of the model all have major uncertain elements. Virtually none of the major components is completely understood. Moreover, because the model embodies long-term projections of poorly understood phenomena, the results should be viewed as having growing error bands the further the projections move into the future.[43]

Nordhaus thus gives us reason to pause before taking his results too literally, especially when we bring genuine uncertainties back into consideration.

Nordhaus claims that his analysis accommodates "the potential for catastrophic consequences from abrupt climate change [because t]hese are included as 'willingness to pay' to avoid the damages that might accompany major climate changes."[44] What this means is that Nordhaus draws on data stemming from individuals' assessments of how much they would be willing to pay to avoid abrupt climate change. I find it hard to understand, however, how Nordhaus thinks it is reasonable for individuals to include, let alone weight, possibilities such as the collapse of the West Antarctic Ice Sheet in willingness-to-pay assessments since as of 2007 (and less explicitly 2013) even the IPCC could not guess the likelihood of this happening or the magnitude of its effects. Nordhaus does claim to account for the fact that climate damages rise non-linearly with temperature increases, but it is not clear how he could include genuinely uncertain but possible outcomes in his analysis.

To be fair, Nordhaus does emphasize that cost-benefit analysis should not be the only guide to decision making. He admits that current models are limited in their ability to look at the potential for catastrophic events, but argues we should not allow fears about low-probability outcomes in the distant future to prevent us from addressing the high-probability dangers we face now.[45] While it is certainly important to address high-probability immediate threats of harm, the problem with Nordhaus's stance is that this cannot justify ignoring distant threats of harm with low or unknown probability. We should not have been ignoring the possibility that the West Antarctic Ice Sheet might start irreversibly melting just because we did not have enough information to assess the likelihood of this scenario. We knew enough to recognize the mechanism by which this could happen and that the conditions for the triggering of this mechanism were accumulating. We certainly should not ignore the fact that it appears this melting is now inevitable just because we do not yet know how quickly this will happen, what all of the consequences of melting will be, and/or because it appears that most of this melting will occur more than 100 (or 200) years from now.

Nordhaus's argument that we should address "clear and present dangers" now and wait to address "unclear and distant threats" later,[46] seems to stem from his argument that:

> The surprising result here is that high-climate-change outcomes are positively correlated with consumption ... Those states in which the global temperature increase is particularly high are also ones in which we are on average richer in the future. This leads to the paradoxical result that there is actually a negative risk premium on high-climate-change outcomes.[47]

What Nordhaus is implying here is not only that future people will be wealthy enough to adapt to climate change but further that because of this there might actually be good economic reasons against mitigating climate change. This point demonstrates the extent to which Nordhaus uses cost-benefit thinking: he is not considering the possibility that increased wealth might not be able to compensate for all damages or harmful effects. The problem with Nordhaus's logic is that we often will not know when it is too late to prevent events that may be catastrophically harmful no matter how much we are willing to spend.

As we have seen, atmospheric greenhouse gas concentrations take a long time to dissipate, which means that there will be a delay between our acting to reduce greenhouse gas emissions and the climatic impacts, driven by global average temperature changes. The longer we delay action to reduce greenhouse gas emissions, the higher global average temperatures are predicted to rise. The problem is that global average temperatures will not immediately decrease if we later decide we want to take action to reduce atmospheric greenhouse gas concentrations. Further, once the West Antarctic Ice Sheet melts, we will not be able to refreeze it or prevent sea levels from rising, no matter how wealthy we are. Nor will future people be able to compensate for five or more meters of sea level rise no matter how wealthy they are because the extent and magnitude of the impacts will simply be too great. Surely there are steps we can take to minimize the harmfulness of sea level rise – hence I argue in the next chapter that adaptation has to play a key role in a precautionary approach to climate change – but wealth will not make up for some losses, such as the inundation of entire nations.

That it now seems very possible that we have already crossed a threshold for the melting of the West Antarctic Ice Sheet has surprised even the scientific community. We should not ignore potentially catastrophic threats about which we know enough to trigger the Catastrophic Precautionary Principle (i.e., where we understand the mechanism by which the threat would be realized and have evidence that the conditions for the function of the mechanism are accumulating) because we might not be able to buy our way out of these events using our increased wealth if we wait too long. Not all harmful outcomes, especially irreversible outcomes, can be compensated for. While of course future people will sometimes be able to compensate for some climate change impacts, for example by developing technologies to prevent such impacts from being harmful, we cannot *assume* that this will always be the case when we are talking about widespread, irreversible changes.

In his 2013 book Nordhaus seems to take more seriously the costs of catastrophic climate impacts, yet his focus remains on a quantitative analysis that struggles to incorporate genuinely uncertain possibilities. He says that while cost-benefit analysis is capable of incorporating tipping points, abrupt changes, discontinuities, and catastrophes, tipping points, for example, are problematic because of "our inability to predict the impacts of the threshold damages reliably."[48] Nordhaus seems to think uncertainties about climate change can be dealt with quantitatively. He argues: "The general point here is that if the damages are uncertain, highly nonlinear, and clifflike in the Climate Casino, then our cost-benefit analysis will generally lower the optimal target to provide insurance against the worst-case options."[49] My worry about Nordhaus inadequately accounting for genuinely uncertain catastrophic outcomes is not resolved by this point, however, since he is talking about tipping points that can be modeled. The real worry is that there are tipping points that are not making it into our models because we do not yet know enough about them. By all means we should work towards doing a better job of modeling the economics of climate change, but given the complexities of the climate system we likely will never be able to count on perfectly modeling climate risks. There is a tension here in that Nordhaus seems to acknowledge this very issue but again ultimately emphasizes the importance of comparing "the price tag with the things we are buying."[50] Unfortunately Nordhaus is not convincing that he does a good enough job calculating those price tags given the uncertainties involved.

Unlike Nordhaus, the *Stern Review* claims to take a precautionary approach to the economics of climate change that attempts to compensate for uncertainty by erring on the side of preventing very bad outcomes. Stern argues that uncertainty provides, "an argument for a more, not less demanding goal, because of the size of the adverse climate-change impacts in the worst-case scenarios."[51] Furthermore, it is also "an argument for setting a more demanding long-term policy, not less, because of the asymmetry between unexpectedly fortunate outcomes and unexpectedly bad ones."[52] The argument here is that we ought to be type II error averse when it comes to dealing with the uncertain outcomes of climate change (erring on the side of being wrong about the worst possible outcomes coming to fruition rather than the best).

This implies that the *Stern Review* looks to the high end of predicted temperatures for a given emissions scenario, but it is not always clear what this means in practice. Again looking back to Figure 4.1, it is at first not clear whether this is akin to looking at the high end of the likely range of predicted temperatures at a given time for an emissions scenario or whether the highest possible modeled temperatures are what should be used. However, Stern himself acknowledges that the PAGE model is "cautious on climate sensitivity" in that its "*full* spread" of model runs "is within the IPCC AR4 '*likely*' (66 percent confidence interval) range."[53] This means that Stern thinks the PAGE model he uses does a good job of considering uncertainty since it looks at the full likely range, but from a broader perspective it is problematic that it does not look at possible outcomes beyond this range. *The Stern Review* therefore cannot avoid the fact that economics has a hard time accounting for low-probability, high-impact possibilities since these are the possibilities that fall outside the likely ranges but are nonetheless important to consider.

The *Stern Review* admits that subjective assessments have to be made about the most extreme possible outcomes of climate change because objective evidence about these risks is limited.[54] Because the consequences of extreme climate change would be so bad and because the *Stern Review* gives equal ethical consideration to future generations, the *Stern Review* argues for erring on the side of preventing such outcomes rather than addressing them when they come up, since by then it will be too late. This is the way in which the *Stern Review* attempts to incorporate precaution directly. The question is whether this approach is sufficiently precautionary. A key worry is that the *Stern Review* does not adequately incorporate the significance of genuinely uncertain outcomes, especially those beyond the likely range.

As Martin Weitzman argues in a 2007 paper, the probability distributions of such outcomes are inherently difficult to estimate.[55] He says:

> In the particular application to the economics of climate change, where there is so obviously limited data and limited information about the global catastrophic reach of climate extremes for the case of $\Delta T > 6°C$, to ignore or suppress the significance of rare tail disasters is to ignore or suppress what economic theory is telling us loudly and clearly is potentially the most important part of the analysis.[56]

Weitzman agrees that we at least need to pay greater attention to the most serious, if unlikely, threats of harm climate change poses. This is the message he and Wagner reiterate in their 2015 book. In his 2007 paper Weitzman identifies the problem as the fact that there is no commonly accepted usable economic framework for extreme tail end disasters.[57] Weitzman ultimately suggests that even the *Stern Review* might not adequately take tail disasters into account in its formal analysis, despite its valiant attempt to do so, precisely because of the lack of an analytical framework for assessing them.[58]

From the perspective of the Catastrophic Precautionary Principle, the hardest-to-quantify climate impacts and damages about which there is much genuine uncertainty are some of the possible outcomes that matter the most. Our best scientific assessments show us that there is a plausible mechanism that could lead to these extremely catastrophic scenarios; even though we do not know exactly what they would look like, we know enough to recognize several degrees of warming will be catastrophically harmful on a wide scale. Even if such outcomes are very unlikely, the Catastrophic Precautionary Principle gives us strong pro tanto reasons to take precautionary measures to try to prevent such outcomes from coming to pass since the consequences would be so dire. So while the *Stern Review* certainly does take a precautionary approach compared to Nordhaus, it is not aggressive enough in its treatment of genuine uncertainty from the precautionary perspective I am defending in this book.

Stern's 2015 book, however, does seem to take more seriously genuinely uncertain yet potentially catastrophic climate impacts. Stern says: "We know that models leave out much that is important – that is what makes them models."[59] He emphasizes that the possibility of extreme climate catastrophes should not be left out of our thinking about climate change and climate policy despite our limited ability to model such outcomes. This contributes to his position that, "the arguments that the costs of inaction greatly exceed the costs of action, strong then [at the time the *Stern Review* was published], are still stronger now."[60] Stern, Wagner, and Weitzman all emphasize that economics should play a crucial role in decision making about climate change while at the same time emphasizing the limits of quantitative modeling. Implicit in Weitzman's 2007 arguments is the idea that once economics is better equipped it will be able to guide decision making.[61] Wagner and Weitzman make this explicit when they say, "[w]e certainly shouldn't scrap economic climate models for their inadequacy. If anything we should be supercharging them ... More manpower and data would at least help the models incorporate the latest available information in real time."[62] That we should continually work to improve our scientific and economic understanding of climate change is certainly a worthy goal. The Catastrophic Precautionary Decision-Making Framework supports such an end too. Yet the acknowledgement that genuine uncertainties do limit – and always will limit – our ability to quantitatively model climate change scientifically and economically reveals a key reason why we should be looking to the Catastrophic Precautionary Principle for guidance.

Uncertainties will always persist in our understanding of climate change, especially about the nature and extent of climate impacts in the far future. The Catastrophic Precautionary Principle is clear about what kinds of threats (i.e., those that could severely harmfully affect many millions of people) and how much we have to know about such threats (i.e., the mechanism by which the threat may be realized and that the conditions for the function of this mechanism are being realized) for us to have strong pro tanto moral reasons to take precautionary measures against the threat. Even Wagner and Weitzman acknowledge ethics comes into play here:

> In the end, it's risk management – existential risk management. And it comes with an ethical component. Precaution is a prudent stance when uncertainties about catastrophic risks are as dominant as they are here. Benefit-cost analysis is important, but it alone may be inadequate, simply because of the fuzziness involved with analyzing high-temperature impacts.[63]

The Catastrophic Precautionary Principle can serve this function and be the ethical principle guiding decision making.

There is a key difference, however, between the Catastrophic Precautionary Principle and Wagner and Weitzman's vision for the role of precaution here. The Catastrophic Precautionary Principle pushes us to recognize even very modest climate change poses threats of catastrophe, whereas Wagner and Weitzman focus their attention on high-warming scenarios. Wagner and Weitzman are right that even low probability (or unknown probability) high-temperature scenarios deserve our attention and warrant precautionary action, but their ultimate commitment to an economic approach means that they ultimately focus on evaluating costs and benefits in much the same way as their economist peers at lower-temperature scenarios (albeit with a lower risk tolerance for the possibility of high-temperature scenarios). Wagner and Weitzman are worried about economic catastrophes in high-warming scenarios, not merely with many millions of people being severely harmfully affected by climate impacts if on balance such outcomes can be compensated for economically. As we will see in the next chapter, the Catastrophic Precautionary Principle supports extremely aggressive mitigation targets – far more aggressive than Wagner and Weitzman push for – as part of a comprehensive precautionary strategy against the diverse threats of climate catastrophes, because it sets a high moral bar.

Stepping back, Stern's assumptions appear precautionary in comparison to those of Nordhaus, but Weitzman's implicit suggestion that Stern's assumptions might not be precautionary enough shows that there may be different degrees to which assumptions are precautionary in an economic analysis. The Catastrophic Precautionary Decision-Making Framework can serve to help decision makers sort out which economic approaches are appropriate for assessing a threat of harm (e.g., one that includes adequate consideration of tail disasters when such possibilities are present), and what information economics has to offer in assessing whether precautionary measures are required and if so what these measures entail. However, we should not lose sight of our moral reasons for taking a precautionary approach to climate change more broadly: we should try to prevent even uncertain threats of catastrophe not (just) because it is economically prudent to do so but because it would be wrong to stand by and let catastrophes unfold. Before the *Stern Review* came out, the implicitly anti-precautionary nature of many economic assessments of climate change may have been harder to recognize. Wagner and Weitzman's work makes the *Stern Review* and even Stern's later analysis seem lacking from a precautionary perspective. Yet the Catastrophic Precautionary Principle is far more demanding in what it calls for than Wagner and Weitzman because of their hyper-focus on extremely high-warming scenarios.

3 Uncertainty as an Argument Against Discounting in Climate Economics

There is another aspect of economic analyses of climate change to which uncertainty is keenly relevant, namely, discounting. Discounting is one of the most contested and problematic aspects in economic analyses of climate change, yet it is also one of the most influential.[64] From a precautionary perspective the reason it is important to talk about discounting is that it too is affected by uncertainty about climate change. In fact, discounting is an inappropriate practice because it masks the very outcomes we should pay the closest attention to, namely those in which many millions (or even billions) of future people may be severely harmfully affected. Uncertainty does not merely justify using a low discount rate, as some like Stern, and Wagner and Weitzman argue. Rather it grounds an argument against discounting altogether. While analyses that use a very low discount rate do a better job from a precautionary perspective of appropriately modeling the economics of climate change, it is also important that we look at all possible consumption streams coming out of economic assessments. Even seemingly precautionary economic assessments may not adequately reveal the full extent of morally relevant climate risks.

The Catastrophic Precautionary Principle gives us reasons to carefully assess whether threats of catastrophe might manifest at any point in the future since, as Daniel Steel says, "[i]n short, a catastrophe is a catastrophe, no matter when it happens."[65] This is where uncertainty becomes relevant to a discussion of discounting in economic assessments of climate change, for there is significant uncertainty about climate sensitivity, especially when we extend out past 2100, the date at which most IPCC graphs end. There is further uncertainty about when and whether catastrophically harmful climate impacts will manifest under different emissions scenarios. Since genuinely uncertain outcomes are not adequately represented (if they are modeled at all), all of the above worries about the implications of uncertainty for the economics of climate change should make us very wary about discounting. It is therefore important that we look at all possible consumption streams, yet discounting prevents this. Discounting is a central part of a procedure in welfare economics intended to represent the impacts of climate change in terms of one summary number (or a range of such numbers). This procedure can problematically mask scenarios in which climate damages cannot be compensated for (e.g., irreversible or massive losses) by aggregating inter-temporal impacts and representing damages in today's dollars. We need to understand what the implications of our policy options are both now and in the future so that we can decide on the most appropriate courses of action.

In order to determine whether threats are catastrophic and in order to determine how we ought to address such threats, we need to know whether and how future generations will be affected and how any threats of catastrophe could be avoided. Discounting has the potential to mask potentially harmful outcomes by not revealing all possible outcomes such that a complete assessment of whether precautionary action is required, and if so what such action entails, is not possible. Robert Lind and Richard Schuler also argue policy makers should see all possible consumption streams, though for very different reasons,[66] so this idea is not unheard of in the economic literature. The Catastrophic Precautionary Decision-Making Framework therefore pushes us – minimally – to evaluate all consumption streams because we must assess whether there might be catastrophically harmful impacts that economic growth will not enable future people to avoid.

Here again Weitzman's, Stern's, and Wagner and Weitzman's perspectives on discounting exemplify debates within the economics of climate change and the difference between even precautionary perspectives and the view I am defending here. The *Stern Review* uses a very low discount rate for analyzing climate change because of the nature of the threats of harm it poses. The *Stern Review* argues that the approach to discounting taken in an economic assessment of climate change must diverge from traditional approaches because such an approach "must meet the challenge of assessing and comparing paths that have very different trajectories and involve very long-term and large inter-generational impacts."[67] Different climate change policies (different emissions scenarios) could lead to drastically different outcomes. The difference between high emissions scenarios and the lowest possible emissions scenarios is a difference of many degrees in terms of global average temperature and hence impacts on humans.

Because of this, the *Stern Review* claims it takes into account ethical arguments in determining an appropriate discount rate. The most influential ethical assumption the *Stern Review* makes is that, "if a future generation will be present, we suppose that it has the same claim on our ethical attention as the current one."[68] The *Stern Review* argues that this implies treating the welfare of future generations as on par with our own because that the prospects of future generations should be taken into account follows from "most standard ethical frameworks."[69] The *Stern Review* here follows the conclusion of John Broome, in the paper he contributed to the work of the *Stern Review*, that he sees "no convincing grounds for discounting future lives."[70] This stance about the ethical standing of future people leads the *Stern Review* to use a very low rate of pure time preference of 0.1%.[71] The reason the *Stern Review* has a positive rate of pure time preference at all is that it accounts for the possibility of human extinction. The *Stern Review* therefore generally uses a discount rate of 1.4%,[72] though it does not always use a consistent discount factor or discount rate.[73] Wagner and Weitzman agree that, "precautionary prudence dictates we should at least consider using low rates for long-term discounting,"[74] suggesting it is hard to defend discount rates above 1 or 2 percent in analyses of climate change.

Nordhaus, on the other hand, consistently uses a relatively large discount rate. In his 2008 analysis he admits that the choice of the discount rate is especially important for climate change because it will significantly impact the future, including the far future. He says: "The approach in the DICE model is to use the estimated market return on capital as the discount rate. The estimated discount rate in the model averages 4 percent per year over the next century."[75] Nordhaus illustrates that according to this rate, this means that US$1,000 of climate damages 100 years from now is valued at $20 today. It is therefore unsurprising that Nordhaus has criticized the *Stern Review*'s "extreme" assumptions economic discounting, which results in its "radical view of policy."[76] He says: "The *Review* seems to have become lost in the discounting trees and failed to see the capital market forest by overlooking the constraints on the two normative parameters."[77] Nordhaus continued to defend his approach to discounting in his 2013 book arguing:

> We need to use a discount rate that reflects the actual market opportunities that societies face, not an abstract definition of equity taken out of the context of market realities. The logic of market discounting is not just a selfish view that the future should take care of itself. It does not hold that we should consume all our income and make no investments to protect our world or future generations. Nor does it hold that we should ignore impacts a few decades in the future. Rather, it reflects that there are many high-yield investments that would improve the quality of life for future generations. The discount rate should be set so that our investable funds are devoted to the most productive uses.[78]

Despite the fact that Nordhaus believes that economic analysis should not be the sole guide to decision making, he clearly has strong opinions about how economic analysis should be done – using a relatively high discount rate.

The biggest problem with Nordhaus's logic in favor of a "traditional" approach to discounting is that climate change does not fit the assumptions made by standard economic approaches insofar as there might be thresholds after which we will not be able to avoid potentially harmful climate impacts, and climate change threatens to have harmful effects that future people might not be able to compensate for. Given that there is so much uncertainty, especially genuine uncertainty, about climate change combined with the possibility of many catastrophic climate damages that may not be compensate-able, discounting may mask important possibilities that should be taken into account when making decisions about climate policy. Stern hence takes a more defensible approach to discounting, though his results too may mask important possible outcomes.

Nordhaus argues that the *Stern Review*'s results stem from its use of a low rate of pure time preference and low inequality aversion between generations.[79] He believes the near-zero pure rate of time preference and correspondingly low discount rate used in the *Stern Review* magnifies "large and speculative damages in the far-distant future" into a large current value.[80] Nordhaus further demonstrates, through a series of basic calculations, that if the *Stern Review*'s parameterizations are corrected to match standard economic methods and assumptions, in part by using a higher discount rate, the results are in line with standard economic models (such as his own).[81] He calculates that more than half of the estimated damages "now and forever" cited by the *Stern Review* occur after the year 2800, which he implies should be taken as reason against the *Review*'s use of an extremely low discount rate.[82] Weitzman estimates that the difference between the *Stern Review*'s discount rate and more traditional discount rates (like that used by Nordhaus) changes the estimated damage costs of climate change 200 years from now by two orders of magnitude.[83]

The fact that it is Nordhaus's and Stern's use of discounting rather than their treatment of uncertainty per se that leads them to such different results and recommendations about climate economics reinforces Wagner and Weitzman's argument that Stern is not as precautionary as he appears. This in turn reinforces the value of the kind of approach recommended by the Catastrophic Precautionary Decision-Making Framework, which requires digging into the details and closely examining how and why various assessments of climate change suggest what they do. In his 2015 book Stern captures something closer to the reasons against discounting I am trying to present here when he emphasizes the importance of examining emissions paths, or more broadly climate policy pathways, by looking at the consequences and risks they bring.[84] Given his emphasis on genuine uncertainties that cannot be modeled, this reasoning provides yet another inroad to recognizing why taking a precautionary approach to climate change makes so much sense. However, it also sheds light on the extent to which the practice of discounting is so entrenched in economic modeling and hence why it is important we have mechanisms like the Catastrophic Precautionary Decision-Making Framework that push us to consider the limitations of all analyses of climate change.

Wagner and Weitzman epitomize the standard view when they suggest that "[t]he one thing we know for sure is that we ought to discount whichever number we get. The basic logic of discounting is sound and ever present."[85] The debate among economists is *what* the discount rate should be, not whether we should be discounting at all. It is important that we look at all possible scenarios and the very practice of discounting masks some of the possible consumption streams that might be the most important to consider. Like Stern, Wagner and Weitzman see uncertainty as a reason to use low rates for long-term discounting rather than as a reason to ensure decision makers have access to all possible consumption streams. One virtue of the Catastrophic Precautionary Decision-Making Framework is that it pushes decision makers not to lean on economic assessments – especially any single economic assessment – too heavily. It can help us understand the limited extent to which Stern is precautionary, the importance of looking at tail disasters, as Wagner and Weitzman emphasize, but also the limitations of standard methodologies like discounting. Considering what the economic costs of climate change might be is important, but uncertainty limits the extent to which we can and should rely on standard economic techniques.

4 Conclusion

Economics is not and cannot be a one-stop-shop for decision makers. Even most theorists using multi-dimensional economic frameworks acknowledge that the complexity of issues like climate change require multi-disciplinary inputs to decision making. Hausman and McPherson identify the appeal and limitations of cost-benefit analysis when they say:

> Policy makers need numbers, and cost-benefit analysis provides numbers. Though the numbers cost-benefit analysis provides are sometimes informative, they are often misleading. There is no substitute for rational deliberation and no uniformly reliable source of information concerning the welfare possibilities or consequences of alternatives.[86]

It is this need for numbers that might drive some to compare cost-benefit analysis to the Catastrophic Precautionary Principle (or mistakenly to "the" precautionary principle) and to prefer the former. However, even seemingly precautionary approaches to the economic analysis of climate change struggle to accommodate uncertainty and can obscure our moral reasons for taking precautionary measures against climate catastrophe. When assessing a temporally diffuse uncertain threat of harm like climate change it is important that economists make appropriate precautionary assumptions, but this alone is not enough since not all precautionary concerns can be adequately modeled or incorporated into economic assessments. Economic assessments of climate change cannot incorporate genuinely uncertain threats of harm, can obscure results in the gray area between risk and uncertainty, and all employ the problematic practice of discounting.

The biggest differences between even Stern or Wagner and Weitzman – economists who take precautionary perspectives – and myself stem from how we think about uncertainty and what we think the aims of climate policy should be. Stern attempts to incorporate a precautionary approach directly into his economic assessment. Wagner and Weitzman go even further with their precautionary perspective because they think that there are good economic reasons to be concerned about tail disasters and climate catastrophes. However, even Wagner and Weitzman's approach is too limited because while I agree we should be worried about the possibility of 6°C of warming (the possibility of which is hidden in the fat tails), we should also be worried about the possibility of 1.2°C of warming. Furthermore, it is critical that we assess all possible outcomes when deciding on appropriately precautionary climate policy and even very limited discounting can mask important possibilities. Precautionary economic assessments do and should play a crucial role in helping us understand the risks of climate change, but ultimately it is ethics that must guide our understanding of unacceptable climate risk. Some economists are incorporating important ethical concerns into their approaches and methodologies, but even as better assessments emerge it is crucial that decision makers use tools like the Catastrophic Precautionary Decision-Making Framework so that they understand how and why any economic assessments they look at arrive at the numbers they do and so that clear moral limits are established.

We have strong pro tanto moral reasons for taking a precautionary approach to threats of catastrophe. We need to turn to the Catastrophic Precautionary Principle and Catastrophic Precautionary Decision-Making Framework to ensure we do not lose sight of this. As Dale Jamieson argues, "[e]conomics alone cannot tell us what to do in the face of climate change. The fundamental problem with climate economics is not that it fails to come up with the right numbers but that there is more at stake than what the numbers reveal."[87] Climate policy should first and foremost aim to satisfy our minimum moral obligations to both existing and future people. The Catastrophic Precautionary Principle can guide us to see that trying to prevent climate change from having catastrophically harmful impacts should be a central aim of climate policy. Economic analyses can provide useful information for decision making about threats of harm, but the Catastrophic Precautionary Decision-Making Framework gives us reason to carefully evaluate the assumptions and methodologies of such analyses when drawing on their results and recommendations. I am not so naïve as to think that all that has been missing from existing climate policies is a commitment to the Catastrophic Precautionary Principle. Global climate policy negotiations are complex and influenced by a huge range of factors. However, a clear, strong precautionary commitment may be an important part of an effective climate regime, which is what I now turn to.

Notes

1 Posner 2005, 140.
2 See also Hammitt 2010; Montgomery and Smith 2010.
3 For other discussions of ethics and the economics of climate change that reach many parallel conclusions to those drawn in this chapter, see Spash 2002; Gardiner 2011; Jamieson 2014; Padilla 2004.
4 Nordhaus 2008, 2007b, 2007a, 2013.
5 Stern 2007, 2015.
6 Wagner and Weitzman 2015; Weitzman 2007.
7 I here follow Knight 2002, 20.
8 For alternative arguments for this conclusion, see Spash 2002; Padilla 2004.
9 Wagner and Weitzman 2015, 37.
10 See Aldred 2012.
11 Thalos 2012, 182.
12 Fleming 1996, 164.
13 Weaver et al. 2013; Lempert et al. 2006; Hall et al. 2012; Kunreuther et al. 2013; Webster et al. 2012.
14 Pindyck 2013.
15 Note that I use this example from the IPCC's AR4, rather than a newer example from AR5, since it helps me make my point in a simpler way (as the emissions scenarios from AR4 are easier to apply to my example than the RCPs used in AR5). This example will also prove useful below since it was current when the debate between Nordhaus, Stern, and Weitzman began.
16 Nordhaus 2008, 6.
17 Nordhaus 2013.
18 Stern 2007, xv.
19 Stern 2015, 150.
20 Wagner and Weitzman 2015, xi.
21 Nordhaus 2008, 8.
22 Nordhaus 2008, 32–33.
23 Nordhaus 2008, 33.
24 Nordhaus 2008, 34.
25 Nordhaus 2008, 5.
26 Nordhaus 2008, xii.
27 Nordhaus 2008, 9.
28 Nordhaus 2008, 80.
29 Stern 2007, ix.
30 Stern 2007, 25.
31 Stern 2007, 28.
32 Stern 2007, 320.
33 Stern 2007, 38.
34 Stern 2007, 1.
35 Stern 2007, 347. At another point the *Stern Review* also says: "Very strong reductions in carbon emissions are required to reduce the risks of climate change. They are likely to provide benefits well in excess of the costs. Indeed the costs of not acting strongly are likely to be very high" (Stern 2007, 640).
36 Stern 2015, chap. 4.
37 Wagner and Weitzman 2015, 77–78.
38 Wagner and Weitzman 2015, 52.
39 Nordhaus 2008, 123.
40 Nordhaus 2008, 36–37.
41 Nordhaus 2007b, 134–37.
42 Nordhaus 2008, 124–25.

43 Nordhaus 2008, 193.
44 Nordhaus 2008, 144.
45 Nordhaus 2008, 147.
46 Nordhaus 2008, 147.
47 Nordhaus 2008, 140.
48 Nordhaus 2013, 214.
49 Nordhaus 2013, 217.
50 Nordhaus 2013, 218.
51 Stern 2007, 318.
52 Stern 2007, 328.
53 Stern 2010, 59.
54 Stern 2007, 327.
55 Weitzman 2007, 723.
56 Weitzman 2007, 719.
57 Weitzman 2007, 723.
58 Weitzman 2007, 722. Eric Neumayer makes a similar argument that the *Stern Review* fails to adequately address the irreversible and non-substitutable damages and loss of natural capital that climate change will incur. He argues that the importance of such damages would have provided an even more compelling case for drastic action to mitigate climate change (Neumayer 2007).
59 Stern 2015, 150.
60 Stern 2015, 303.
61 Weitzman does connect his argument to "the" precautionary principle, though he does not appear to endorse a precautionary principle separate from the economic advances for which he argues: "The general point is that from experience alone one cannot acquire sufficiently accurate information about the probabilities of tail disasters to prevent the expected marginal utility of an extra sure unit of consumption from becoming unbounded for *any* utility function having everywhere-positive relative risk aversion, thereby effortlessly driving cost-benefit applications of expected utility theory. The degree to which this kind of 'generalized precautionary principle' is relevant in a particular application must be decided on a case-by-case basis that depends upon the extent to which a priori knowledge in a particular case limits the extent of posterior-predictive tail thickening" (Weitzman 2007, 719).
62 Wagner and Weitzman 2015, 60.
63 Wagner and Weitzman 2015, 78.
64 For an in-depth review of the literature on this topic see Davidson 2015. See also Kolstad et al. 2014.
65 Steel 2013, 335.
66 Lind and Schuler 1998.
67 Stern 2007, 25.
68 Stern 2007, 35.
69 Stern 2007, 37.
70 Broome 2006, 19. For a related argument that a human rights-based account requires that we do not have any pure time preference, see Caney 2008, 2009. For the provocative claim that much disagreement over pure time preference stems from unstated differences concerning the very aims of welfare economics, see Kelleher, n.d.
71 Stern 2007, 663.
72 As described in Ackerman and Stanton 2015, 69; Weitzman 2007; Dasgupta 2008.
73 Stern 2007, chap. 2A.
74 Wagner and Weitzman 2015, 69.
75 Nordhaus 2008, 10, see also his chapter 9.

76 Nordhaus 2007b, 689.
77 Nordhaus 2007b, 700.
78 Nordhaus 2013, 193.
79 Nordhaus 2007b, 202.
80 Nordhaus 2007b, 696.
81 Nordhaus 2007b, 697–701.
82 Nordhaus 2007b, 696.
83 Weitzman 2007, 708.
84 Stern 2015, chap. 5.
85 Wagner and Weitzman 2015, 68.
86 Hausman and McPherson 2006, 154.
87 Jamieson 2014, 9.

References

Ackerman, Frank, and Elizabeth A. Stanton. 2015. *Climate Change and Global Equity*. New York: Anthem Press.

Aldred, J. 2012. "Climate Change Uncertainty, Irreversibility and the Precautionary Principle." *Cambridge Journal of Economics* 36(5): 1051–1072. doi:10.1093/cje/bes029.

Broome, John. 2006. "Valuing Policies in Response to Climate Change: Some Ethical Issues." http://webarchive.nationalarchives.gov.uk/20130129110402/http://www.hm-treasury.gov.uk/d/stern_review_supporting_technical_material_john_broome_261006.pdf.

Caney, Simon. 2008. "Human Rights, Climate Change, and Discounting." *Environmental Politics* 17(4): 536–555. doi:10.1080/09644010802193401.

Caney, Simon. 2009. "Climate Change and the Future: Discounting for Time, Wealth, and Risk." *Journal of Social Philosophy* 40(2): 163–186. doi:10.1111/j.1467-9833.2009.01445.x.

Dasgupta, Partha. 2008. "Discounting Climate Change." *Journal of Risk and Uncertainty* 37(2–3): 141–169. doi:10.1007/s11166-008-9049-6.

Davidson, Marc D. 2015. "Climate Change and the Ethics of Discounting." *Wiley Interdisciplinary Reviews: Climate Change* 6(4): 401–412. doi:10.1002/wcc.347.

Fleming, David. 1996. "The Economics of Taking Care: An Evaluation of the Precautionary Principle." In *The Precautionary Principle and International Law: The Challenge of Implementation*, edited by David Freestone and Ellen Hey, 147–167. London: Kluwer Law International.

Gardiner, Stephen M. 2011. *A Perfect Moral Storm: The Ethical Tragedy of Climate Change*. Oxford: Oxford University Press.

Hall, Jim W., Robert J. Lempert, Klaus Keller, Andrew Hackbarth, Christophe Mijere, and David J. McInerney. 2012. "Robust Climate Policies under Uncertainty: A Comparison of Robust Decision Making and Info-Gap Methods." *Risk Analysis: An Official Publication of the Society for Risk Analysis* 32(10): 1657–1672. doi:10.1111/j.1539-6924.2012.01802.x.

Hammitt, James K. 2010. "Global Climate Change: Benefit-Cost Analysis vs. the Precautionary Principle." *Human and Ecological Risk Assessment: An International Journal* 6(3): 387–398. doi:10.1080/10807030091124536.

Hausman, Daniel M., and Michael S. McPherson. 2006. *Economic Analysis, Moral Philosophy, and Public Policy*, second edition. Cambridge: Cambridge University Press.

IPCC. 2007. "Summary for Policymakers." In *Climate Change 2007: The Physical Science Basis. Contribution of Working Group I to the Fourth Assessment Report of the Intergovernmental Panel on Climate Change*, ed. S. Solomon, D. Qin, M. Manning, Z. Chen, M. Marquis, K.B. Averyt, M. Tignor, and H.L. Miller. Cambridge: Cambridge University Press.

Jamieson, Dale. 2014. *Reason in a Dark Time: Why the Struggle Against Climate Change Failed – and What it Means for Our Future.* Oxford: Oxford University Press.

Kelleher, J. Paul. n.d. "Pure Time Preference in Intertemporal Welfare Economics." Forthcoming at *Economics and Philosophy.*

Knight, Frank Hyneman. 2002. *Risk, Uncertainty and Profit.* Beard Books.

Kolstad, C., K. Urama, J. Broome, A. Bruvoll, M. Cariño Olvera, D. Fullerton, C. Gollier, et al. 2014. "2014: Social, Economic and Ethical Concepts and Methods." In *Climate Change 2014: Mitigation of Climate Change. Contribution of Working Group III to the Fifth Assessment Report of the Intergovernmental Panel on Climate Change*, ed. O. Edenhofer, R. Pichs-Madruga, Y. Sokona, E. Farahani, S. Kadner, K. Seyboth, A. Adler, et al., 1465. Cambridge: Cambridge University Press.

Kunreuther, Howard, Geoffrey Heal, Myles Allen, Ottmar Edenhofer, Christopher B. Field, and Gary Yohe. 2013. "Risk Management and Climate Change." *Nature Climate Change* 3(5): 447–450. doi:10.1038/nclimate1740.

Lempert, Robert J., David G. Groves, Steven W. Popper, and Steve C. Bankes. 2006. "A General, Analytic Method for Generating Robust Strategies and Narrative Scenarios." *Management Science* 52(4): 514–528. doi:10.1287/mnsc.1050.0472.

Lind, Robert C., and Richard E. Schuler. 1998. "Equity and Discounting in Climate-Change Decisions." In *Economics and Policy Issues in Climate Change*, ed. William D. Nordhaus, 59–96. Resources for the Future Press.

Montgomery, W. David, and Anne E. Smith. 2010. "Global Climate Change and the Precautionary Principle." *Human and Ecological Risk Assessment: An International Journal* 6(3): 399–412. doi:10.1080/10807030091124545.

Neumayer, Eric. 2007. "A Missed Opportunity: The Stern Review on Climate Change Fails to Tackle the Issue of Non-Substitutable Loss of Natural Capital." *Global Environmental Change* 17(3–4): 297–301. doi:10.1016/j.gloenvcha.2007. 04. 001.

Nordhaus, William. 2007a. "Economics. Critical Assumptions in the Stern Review on Climate Change." *Science (New York, N.Y.)* 317(5835): 201–202. doi:10.1126/science.1137316.

Nordhaus, William D. 2007b. "A Review of the Stern Review on the Economics of Climate Change." *Journal of Economic Literature* 45(3): 686–702. doi:10.1257/jel.45.3.686.

Nordhaus, William. 2008. *A Question of Balance: Weighing the Options on Global Warming Policies.* New Haven: Yale University Press.

Nordhaus, William. 2013. *The Climate Casino: Risk, Uncertainty, and Economics for a Warming World.* New Haven: Yale University Press.

Padilla, Emilio. 2004. "Climate Change, Economic Analysis and Sustainable Development." *Environmental Values* 13(4): 523–544. doi:10.3197/096327104272622.

Pindyck, Robert S. 2013. "The Climate Policy Dilemma." *Review of Environmental Economics and Policy* 7(2): 219–237. doi:10.1093/reep/ret007.

Posner, Richard A. 2005. *Catastrophe: Risk and Response*. Oxford: Oxford University Press.

Spash, Clive L. 2002. *Greenhouse Economics: Value and Ethics*. London: Routledge.

Steel, Daniel. 2013. "The Precautionary Principle and the Dilemma Objection." *Ethics Policy Environment* 16(3): 321–340.

Stern, Nicholas. 2007. *The Economics of Climate Change: The Stern Review*. Cambridge: Cambridge University Press.

Stern, Nicholas. 2010. "The Economics of Climate Change." In *Climate Ethics: Essential Readings*, ed. Stephen M. Gardiner, Simon Caney, Dale Jamieson, and Henry Shue, 39–76. Oxford: Oxford University Press.

Stern, Nicholas. 2015. *Why are We Waiting? The Logic, Urgency, and Promise of Tackling Climate Change*. Cambridge: MIT Press.

Thalos, Mariam. 2012. "Precaution has its Reasons." In *The Environment: Philosophy, Science and Ethics (Topics in Contemporary Philosophy)*, ed. William P. Kabasenche, Michael O'Rourke, and Matthew H. Slater, 171–184. Cambridge: MIT Press.

Wagner, Gernot, and Martin L. Weitzman. 2015. *Climate Shock: The Economic Consequences of a Hotter Planet*. Princeton: Princeton University Press.

Weaver, Christopher P., Robert J. Lempert, Casey Brown, John A. Hall, David Revell, and Daniel Sarewitz. 2013. "Improving the Contribution of Climate Model Information to Decision Making: The Value and Demands of Robust Decision Frameworks." *Wiley Interdisciplinary Reviews: Climate Change* 4(1): 39–60. doi:10.1002/wcc.202.

Webster, Mort, Andrei P. Sokolov, John M. Reilly, Chris E. Forest, Sergey Paltsev, Adam Schlosser, Chien Wang, et al. 2012. "Analysis of Climate Policy Targets under Uncertainty." *Climatic Change* 112(3–4): 569–583. doi:10.1007/s10584-011-0260-0.

Weitzman, Martin L. 2007. "A Review of the Stern Review on the Economics of Climate Change." *Journal of Economic Literature* 45(3): 703–724. doi:10.1257/jel.45.3.703.

5 Responding to the Threat of Climate Catastrophe

Taking a precautionary approach to climate policy is in one way simple – try to prevent climate catastrophes! – and yet it is also very complex. The Catastrophic Precautionary Principle is a useful guide to climate policy because it gives us a clear benchmark for identifying unacceptable climate risks: precautionary climate policy should minimally aim to prevent climate catastrophes now and into the future. This chapter explores how the Catastrophic Precautionary Principle may be incorporated into global climate policy to help ground aggressive climate policy on all fronts (mitigation, adaptation, and geoengineering). The Catastrophic Precautionary Decision-Making Framework can serve as our guide to determining an appropriate precautionary strategy against climate catastrophe(s). It can help us identify both the strongly action-guiding recommendations that follow from the Catastrophic Precautionary Principle for climate policy and the places where other considerations will have to come into play before we can determine how to implement the precautionary measures recommended by this approach.

Now more than ever we must accept that we cannot avoid climate catastrophe(s) through mitigation alone; adaptation will also have to be an important part of a precautionary strategy for avoiding climate catastrophe(s). The IPCC's assessment of climate change not only suggests climate change already threatens to be catastrophically harmful but also that the best way to reduce the chance of crossing thresholds beyond which further catastrophically harmful climate impacts will become inevitable is to stabilize atmospheric GHGs as low as possible. Clearly we have failed to appropriately mitigate climate change from a precautionary perspective, but it is still in our power to stabilize atmospheric concentrations of GHGs at levels that minimize climate risk. This implies that we have strong pro tanto moral reasons to reduce GHG emissions as quickly as possible, but it also implies that we ought to be implementing precautionary strategies for minimizing the harmfulness of climate impacts that are already occurring or are locked in by warming already in the pipeline. However, because we have so far failed to mitigate climate change meaningfully, some people have begun to question whether there are additional solutions or stopgaps to the climate problem via geoengineering strategies, which aim to intentionally manipulate the global climate.[1] Since the Catastrophic Precautionary Decision-Making Framework guides us to consider all available precautionary measures, we should assess whether geoengineering strategies should be a part of our suite of precautionary measures against climate change. Here too the Catastrophic Precautionary Principle provides guidance by setting a clear boundary on acceptable risks, namely precautionary measures should not themselves create or exacerbate threats of catastrophe.

Before getting into the details of how the views presented in this book weigh in on mitigation, adaptation, and geoengineering specifically, I begin in the first section with a more general discussion of how the Catastrophic Precautionary Principle can guide climate policy. The remaining sections specify what the Catastrophic Precautionary Principle implies about mitigation, adaptation, and geoengineering and how it weighs in on debates in these areas. In each case I carefully identify the powerful implications of the Catastrophic Precautionary Principle (e.g., that we should be taking an extremely aggressive approach to mitigation that is far more ambitious than a 2°C goal), as well as those areas where other moral and political principles will need to come into play to work out how we should implement our precautionary obligations (e.g., determining who should fund adaptation measures on a global and local scale). The key conclusions and recommendations of this chapter are:

- Integrating the Catastrophic Precautionary Principle into guiding climate policies would define a minimally acceptable precautionary approach to addressing climate change (i.e., that the aim of such policies should be to prevent many millions of people from being severely harmfully affected by climate change).
- Mitigation efforts should aim to stabilize atmospheric CO_2 and other GHGs as low as possible. A 1.5° or 2°C stabilization goal will not sufficiently mitigate the risk of catastrophic climate change.
- Adaptation efforts should be globally coordinated so as to "fill the mitigation gap" and prevent many millions of people from being severely harmfully affected by climate impacts.
- Research into geoengineering strategies as possible complements to mitigation and adaptation should continue, but any geoengineering strategy that threatens to create or exacerbate a threat of catastrophe should not be considered as a morally appropriate precautionary measure against climate change.

1 Precautionary Climate Policy

The Catastrophic Precautionary Principle has the potential to be a meaningful driver of precautionary climate policy because it sets clear boundaries around morally unacceptable climate change. It provides strong reasons for taking precautionary measures against climate catastrophes in a way that opens the door to diverse strategies for addressing climate change in comprehensive climate policies. A guiding precautionary climate policy aimed at preventing catastrophic outcomes may look something like the following:

Guiding Precautionary Climate Policy

The aim of climate policy shall be to prevent catastrophic climate change per the Catastrophic Precautionary Principle, where catastrophes are understood to involve many millions of people experiencing severely harmful outcomes. In order to discern what the Catastrophic Precautionary Principle requires, parties will appeal to the Catastrophic Precautionary Decision-Making Framework when assessing climate change impacts and determining further action-guiding precautionary climate policies.

While simple, this kind of guiding policy would have built into it the full force of the Catastrophic Precautionary Principle and set clear limits on acceptable climate risk. Full statements of the Catastrophic Precautionary Principle and Catastrophic Precautionary Decision-Making Framework would have to be provided to ensure what this entails is clear because of all that is inherent to them (e.g., clauses about what constitutes severe harm and the importance of taking immediate precautionary measures against imminent threats). It could ground a new global climate policy or be integrated into existing frameworks such as the UNFCCC.

The UNFCCC is often cited as already containing a version of "the" precautionary principle, but the precautionary approach it specifies is much weaker than that captured by the Catastrophic Precautionary Principle. Looking at articles two and three of the UNFCCC reveals both the extent to which the UNFCCC already espouses taking a precautionary approach to climate policy and the ways in which incorporating the Catastrophic Precautionary Principle could significantly strengthen its precautionary commitments.

United Nations Framework Convention on Climate Change (1992)

ARTICLE 2

OBJECTIVE: The ultimate objective of this Convention and any related legal instruments that the Conference of the Parties may adopt is to achieve, in accordance with the relevant provisions of the Convention, stabilization of greenhouse gas concentrations in the atmosphere at a level that would prevent dangerous anthropogenic interference with the climate system. Such a level should be achieved within a time-frame sufficient to allow ecosystems to adapt naturally to climate change, to ensure that food production is not threatened and to enable economic development to proceed in a sustainable manner.

ARTICLE 3

PRINCIPLES: In their actions to achieve the objective of the Convention and to implement its provisions, the Parties shall be guided, inter alia, by the following:

1   The Parties should protect the climate system for the benefit of present and future generations of humankind, on the basis of equity and in accordance with their common but differentiated responsibilities and respective capabilities. Accordingly, the developed country Parties should take the lead in combating climate change and the adverse effects thereof.
2   The specific needs and special circumstances of developing country Parties, especially those that are particularly vulnerable to the adverse effects of climate change, and of those Parties, especially developing country Parties, that would have to bear a disproportionate or abnormal burden under the Convention, should be given full consideration.
3   The Parties should take precautionary measures to anticipate, prevent or minimize the causes of climate change and mitigate its adverse effects. Where there are threats of serious or irreversible damage, lack of full scientific certainty should not be used as a reason for postponing such measures, taking into account that policies and measures to deal with climate change should be cost-effective so as to ensure global benefits at the lowest possible cost. To achieve this, such policies and measures should take into account different socio-economic contexts, be comprehensive, cover all relevant sources, sinks and reservoirs of greenhouse gases and adaptation, and comprise all economic sectors. Efforts to address climate change may be carried out cooperatively by interested Parties.
4   The Parties have a right to, and should, promote sustainable development. Policies and measures to protect the climate system against human-induced change should be appropriate for the specific conditions of each Party and should be integrated with national development programmes, taking into account that economic development is essential for adopting measures to address climate change.
5   The Parties should cooperate to promote a supportive and open international economic system that would lead to sustainable economic growth and development in all Parties, particularly developing country Parties, thus enabling them better to address the problems of climate change. Measures taken to combat

climate change, including unilateral ones, should not constitute a
means of arbitrary or unjustifiable discrimination or a disguised
restriction on international trade.[2]

The first thing to note is that principle three in the third article is often
referred to as a version of the precautionary principle. This is not surprising
since it explicitly calls for precautionary measures against climate change
that both try to minimize the extent of climate impacts and mitigate the
adverse effects of eventual climatic changes and that uncertainty cannot be
cited as a reason to delay precautionary action. So while article two is
framed in terms of mitigation insofar as the stated goal of the UNFCCC
is to prevent dangerous climate change, principle three expands the range
of precautionary measures suggested by the UNFCCC. Read in the context
of article two, principle two seems to suggest diverse precautionary measures
should be taken to prevent dangerous climate change. This is where the
language in the UNFCCC connects to concepts applied by the IPCC, for
the IPCC uses the concept of dangerous climate change in its framing of
the reasons for concern and key risks posed by climate change.

In this sense there is a clear connection between the framing of the
UNFCCC and the Catastrophic Precautionary Principle, as both seem to
suggest dangerous climate change should be avoided. However, this also
reveals a key way in which the Catastrophic Precautionary Principle is much
more demanding than the precautionary stance built into the UNFCCC
because the concept of "dangerous climate change" is notoriously vague.
For example, there have been vastly different interpretations of what this
implies about what our mitigation targets should be. The Catastrophic
Precautionary Principle is much more explicit, however, in that it sets a
clear floor: if millions of people are at risk of being severely harmfully
affected by climate change in general or by a specific impact we ought to
take precautionary measures to avert catastrophe. If integrated into
the UNFCCC, the Catastrophic Precautionary Principle would hence set
a minimal aim – to prevent catastrophic outcomes – that is more defensible
and explicitly defined than the current stated aim of avoiding dangerous
climate change.

Another way in which principle three is weaker than the Catastrophic
Precautionary Principle is evident in the wording of the second sentence of
this principle. The UNFCCC specifies that scientific uncertainty should
not be used as a reason for postponing precautionary measures against
threats of serious or irreversible damage, which, while valid, is not as
strong as it could be. A precautionary approach should push us not just to
avoid using scientific uncertainty as a reason for inaction but also to
recognize that certain threats demand action so long as a minimum standard
of evidence is met. The Catastrophic Precautionary Principle does this
clearly and explicitly by detailing what this standard of evidence is and
emphasizing that irreversible threats of harm merit extra consideration

because of their potential to be catastrophically harmful in the long run. Namely, it requires that we understand the mechanism by which a threat would be realized and have evidence that the conditions for the function of this mechanism are accumulating. The UNFCCC suggests a precautionary approach by mentioning the issue of scientific uncertainty, but replacing this with a commitment to the Catastrophic Precautionary Principle would make the demands of the UNFCCC with respect to uncertainty far more explicit and demanding.

In this same sentence there is a clause about cost-effectiveness. While I have repeatedly emphasized that economic considerations should be left out of a pro tanto moral principle like the Catastrophic Precautionary Principle, the UNFCCC is actually quite careful in how these economic constraints are framed. Unlike some so-called versions of the precautionary principle that put a limit on the absolute costs of precautionary measures, principle three of the UNFCCC merely requires that precautionary measures be cost effective, and this seems like a reasonable requirement. The UNFCCC could simultaneously commit to the Catastrophic Precautionary Principle and to ensuring precautionary measures are cost-effective so long as moral considerations are the primary drivers in the selection of an appropriate precautionary strategy for addressing threats of climate catastrophe. Principle five also addresses economic considerations but not to the exclusion of the kinds of moral considerations the Catastrophic Precautionary Principle could bring to the UNFCCC. As suggested above, economic impacts can be severely harmful. So promoting climate policies that support the global economic system in ways that will prevent indirectly harmful outcomes is certainly a reasonable goal. In fact, taken as a whole, the principles in article three do a great job of articulating the key competing demands that will and should come into play in setting precautionary climate policies. For example, principle four emphasizes the importance of promoting sustainable development because of its connection to climate change. Implicit here is the notion that whenever possible, broader social justice goals should be integrated into climate policy. There is no reason not to seek out ways to promote sustainable development in ways that will serve multiple goals in the name of addressing climate change – the Catastrophic Precautionary Decision-Making Framework also pushes us to do just this.

Principle one makes it clear that the UNFCCC is already focused on protecting the well-being of humans, that it takes an anthropocentric perspective. So in this way, adding a commitment to the Catastrophic Precautionary Principle would not reorient the foundational aims of the UNFCCC. Rather, it would push the UNFCCC to have a more aggressive precautionary agenda. On the other hand, the wording of article two limits the applicability of the UNFCCC to anthropogenic (as opposed to naturally occurring) climate change. While in practice this does not make a difference, the Catastrophic Precautionary Principle would not stress this point since it pushes us to address all threats of catastrophe, whatever their cause. At the same time, the causal complexity of climate change may lend itself to this more neutral approach. The Catastrophic Precautionary Principle avoids entanglement in arguments over causation, though who or what is responsible for causing climate change will certainly come into play when determining who has responsibility to take and pay for precautionary measures.

In sum, incorporating the Catastrophic Precautionary Principle into the UNFCCC would strengthen the precautionary approach it calls for. For the same reasons that the Catastrophic Precautionary Principle is easy to accept – we at least ought to try to avoid those outcomes that could be catastrophically harmful – the UNFCCC should commit to this principle. Doing so would not only make commitments already implicit to the UNFCCC more concrete but would also set a stronger minimum bar for adequate global climate policy. Granted, the reason the UNFCCC is limited in the way it is likely stems from the same political inertia that has prevented sufficiently precautionary mitigation policies from being adopted and implemented. Nonetheless, the Catastrophic Precautionary Principle would be a powerful means of clearly articulating a more concrete precautionary goal. The remaining sections will explore the more specific implications of this for mitigation, adaptation, and geoengineering.

2 A Precautionary Approach to Mitigation – Against the 1.5/2°C Target

Aggressive mitigation is the most obvious part of a precautionary strategy against climate catastrophe. Mitigation measures work to reduce the causal forces contributing to climate change, primarily GHG emissions, though mitigation also includes things like land use change that can indirectly emit GHGs. Mitigation can take many forms from directly reducing fossil fuel consumption to promoting low- or no-meat diets, since animal agriculture processes significantly more GHG emissions than plant-based agriculture. Given how likely it is climate change will be catastrophically harmful even if we significantly reduce GHG emissions and the range of thresholds beyond which a range of catastrophically harmful climate impacts become imminent, the Catastrophic Precautionary Principle provides very strong reasons for implementing a very aggressive mitigation policy. It cannot be our only guide to mitigation, but, as we will see, it can provide a solid foundation from which the details of a mitigation policy could be worked out.

There is growing consensus that our climate change mitigation goal should be to prevent average surface temperatures from rising more than 1.5°C or 2°C above preindustrial levels. The focus of much recent work in economics, for example, is on the feasibility of meeting a 2°C mitigation target.[3] The UNFCCC's 2015 Paris Agreement aims to hold "the increase in global average surface temperature to well below 2°C above preindustrial levels" and pursue "efforts to limit the temperature increase to 1.5°C above preindustrial levels."[4] Yet the global community has been so focused on 2°C that this new aim of 1.5°C will require a refocusing of research priorities.[5] While the Paris Agreement might seem like a big step forward for global climate policy, from a precautionary perspective there are a number of problems with both the 1.5°C and 2°C goals (hereafter collectively referred to as the 1.5/2°C target). First and foremost, even a 1.5°C mitigation goal is not ambitious enough. While the risks of exceeding either 1.5°C or 2°C global average surface temperatures are clear and dire, there are already significant risks of catastrophic harm from even the seemingly modest temperature increases we have experienced to date. The precautionary perspective supported by the Catastrophic Precautionary Principles gives us strong pro tanto moral reasons to take a much more ambitious mitigation strategy than the 1.5/2°C target suggests. However, there are two other precautionary reasons that relate to moral corruption and the risk of masking uncertainty for being skeptical of the 1.5/2°C target that should push us towards embracing a less specific, albeit much more ambitious goal.

2.1 Why Our Moral Reasons Matter – The Risk of Moral Corruption

The first concern is that overemphasizing a specific temperature target detracts from the moral reasons that motivated setting a mitigation target in the first place. The 1.5/2°C target is driven by concerns about dangerous climate change, which as we have seen can be understood as being grounded in a precautionary approach. The Paris Agreement may seem like a step forward for global climate policy, but there are good reasons to be wary of embracing the 1.5/2°C target too tightly. For to do so is to imply that the 1.5/2°C target is *good enough* when from a precautionary perspective it is not nearly good enough, it is a far cry from where we should be or where we should be aiming. Since we have strong moral reasons to take aggressive precautionary measures against threats of catastrophe, to endorse a mitigation target of 1.5/2°C as good enough is to accept morally inappropriate risks for humanity now and into the future.

1.5/2°C might seem good enough *to us*, the current generation, but from a timeless perspective the risks of this mitigation pathway are simply too steep. We owe it to future generations to take more aggressive measures since we know that climate change threatens to be catastrophically harmful if we do not. It matters that we focus on the right reasons for mitigating climate change because hyper-focusing on a specific temperature target without remembering *why* this target was set in the first place may cause us to fail in our real goal: avoiding dangerous/catastrophic climate change. If we set ourselves on a path to stabilize global average surface temperature at 1.5 or 2°C it may become even harder (if not impossible) to achieve a more ambitious goal.

One way to understand the core worry here is through the lens of what Stephen Gardiner has called the challenge of moral corruption.[6] Gardiner argues that climate change presents a perfect moral storm that threatens our ability to act ethically. Because climate change is both global and intergenerational in reach, while also presenting all of the hardest kinds of theoretical and philosophical challenges (e.g., raises questions about humanity's relationship to the rest of nature and about the treatment of risk and uncertainty), it is extremely hard even to understand what it would mean to respond to these challenges in an ethical way, let alone actually act ethically. What is likely to happen in such a context, argues Gardiner, is that we become morally corrupt, which can manifest in any number of ways. However, the idea is that we might, for example, deceive ourselves into thinking that we are acting ethically when in fact we are ignoring an important part of the ethical challenge of climate change in a way that undermines the ethicality of our actions.

This means that in fighting to realize the 1.5/2°C target we may *think* we are acting ethically when we are ignoring – consciously or unconsciously – the fact that from an intergenerational perspective this target is not ambitious enough and hence is unethical. It might be tempting, for example, to think working towards the 1.5/2°C target is a good first step towards more ethical global climate policy, but there is a risk that by embracing this target we may make it impossible to realize our moral obligation to take aggressive precautionary measures against all threats of climate catastrophe. To embrace the 1.5/2°C target as ethical, therefore, is to deceive ourselves and fail to act on our strong moral reasons to fight for a much more aggressive target. This is why it matters that we remember why we ought to mitigate climate change in the first place: because it is bad that climate change has the potential to be severely harmful to many, many millions of people. At the very least we ought to try to prevent catastrophically harmful outcomes, which is why we should adopt the Catastrophic Precautionary Principle as a key foundation to climate policy.

Some might respond to this line of argument by saying, "but isn't the Paris Agreement and the 1.5/2°C target a good step forward?" The answer is that no, it very well may not be for two reasons. For if it is distracting us from what we really ought to be doing and, worse, is making the possibility of doing what we really ought to be doing that much harder (if not impossible), then this is not really a step forwards at all. Rather, it is a step sideways or even backwards. Of course, part of the worry here is that given the way things actually are in the world, our global political reality, we simply cannot hope for more. So we might accept that the 1.5/2°C target is not morally ambitious enough while embracing it as the best we can aim for politically. The first problem with this notion is that this does not seem to be what is happening with the Paris Agreement. The language in the agreement and that is used to describe the agreement suggests that the 1.5/2°C target is where we should in fact be aiming, that this will help us prevent dangerous climate change. This is exactly the kind of moral corruption I think Gardiner has in mind: we are deluding ourselves into thinking limiting climate change to 1.5/2°C is good enough when we ought to be embracing a much more aggressive goal. The second problem with the "best we can do for now" mentality is that a 1.5/2°C target may very well prevent us from doing better, doing what we ought to be doing.

It is important that we not lose sight of the moral reasons motivating our mitigation efforts in the first place for to do so may lead us to embrace morally inappropriate courses of action, as is happening with the 1.5/2°C mitigation target. While the Catastrophic Precautionary Principle is certainly not a magic bullet against the political inertia plaguing global climate policy, it could provide the strong moral grounding needed to ensure we set an appropriate mitigation target. While a 1.5/2°C goal may seem daunting enough, we should not let our own worries about the feasibility of a more aggressive target get in the way of our fighting to get the world on a more just path. Thinking about the complex ways in which climate change threatens to be harmful across generations should be enough to help us see that our own concerns are minor in the face of the significant catastrophes that may be realized across generations if we fail to act. To meet the demands of morality we must push the world to change. Failing to do so inflicts grave risks of catastrophic climate change on people in the near and distant future. We should be taking a precautionary approach to mitigation policy that is in part guided by the Catastrophic Precautionary Principle as this will help us avoid the morally corrupt temptation of thinking a 1.5/2°C target is good enough.

2.2 The Risk of Masking Uncertainty

The second issue with the 1.5/2°C target is that framing a mitigation goal in terms of a temperature target masks a range of issues about risk and uncertainty. How sure should we be that our actions will set us on a path to avoid the 1.5/2°C temperature target? How confident do we have to be that a given emissions pathway will keep us below the target? Is a 50% chance of staying below the target good enough? Framing mitigation goals as temperature targets masks relevant uncertainties. Whereas climate policies like the Paris Agreement tend to emphasize target stabilization temperatures, scientists are increasingly focusing on atmospheric GHG concentration stabilization targets or IPCC RCPs. The advantage of framing climate policy in terms of a temperature target is that this allows us to think in terms of the climate impacts that are predicted to occur at different temperatures, but the disadvantage of this approach is that we cannot precisely work backwards and determine what level of atmospheric GHGs, let alone GHG emissions, would lead to a given temperature target. Uncertainty about how the climate will respond to atmospheric concentrations of GHGs (captured by the notion of climate sensitivity) means that at best we can aim to avoid crossing a temperature threshold with some specified degree of likelihood (say, 50%). However, this uncertainty can be masked by a stated temperature target.

Framing our mitigation goal in terms of a concentration target, on the other hand, helps make the uncertainty associated with this goal more explicit because it forces us to look at the range of possible temperature stabilizations associated with a particular GHG stabilization goal. At the same time, since the real aim of climate policy should be to avoid crossing thresholds beyond which climate change will be severely harmful, Timothy Lenton has argued that climate policy should be framed to limit anthropogenic radiative forcing, which is the cumulative result of the human drivers of climate change.[7] The IPCC RCPs fall somewhere in between these latter two approaches, in that each concentration pathway represents a trajectory to a long-term GHG concentration and radiative forcing outcome.[8] The special emphasis here is that it matters not only *what* our long-term mitigation targets are but also *how we get there*. The lesson for a precautionary approach to avoiding catastrophic risks is that different mitigation target framings invite and/or mask different uncertainties.

So what mitigation goal does the Catastrophic Precautionary Principle support? It gives us strong moral reasons to avoid any pathway that involves a risk of catastrophe. Paying close attention to the ways in which uncertainty enters into our understanding of such threats is central to the precautionary approach it captures. Since we are already on track to see potentially catastrophically harmful outcomes, we ought to be mitigating climate change to the greatest extent possible given other moral demands. Unfortunately it is hard to put a number on what this would look like. Fortunately, though, many scientists, most notably James Hansen, are concerned about the threat of climate catastrophe in ways that cohere with the Catastrophic Precautionary Principle's definition of catastrophic harm.

Their scientific assessments have led them to advocate for limiting the concentration of CO_2 in the atmosphere to no more than 350 ppm by volume (and possibly even less).[9] Currently, however, the atmospheric CO_2 concentration exceeds 400 ppm.[10] Beyond 350 ppm, there is an increased risk of irreversible climatic changes such as abrupt shifts in forest and agricultural systems and the loss of major ice sheets; hence stabilizing CO_2 above this level risks crossing key climate thresholds, which may be a key planetary boundary.[11] (Of course, CO_2 is not the only relevant GHG, but it is a primary driver of climate change especially from an intergenerational perspective because of its long lifespan in the atmosphere. Other GHG emissions such as methane should certainly be reduced as part of a comprehensive mitigation strategy as well.) Recent data about ice sheet melting and the associated risks from sea level rise have led Hansen to argue that the situation is quite urgent.[12] This analysis suggests we must limit cumulative fossil fuel emissions to 500 gigatons of carbon (GtC) (and find ways to store at least 100 GtC in the biosphere and soil). Without rapid emissions reductions the risk of catastrophic climate impacts is significantly higher.[13] While the Catastrophic Precautionary Principle provides moral reasons to support Hansen's and others' scientific assessments that we ought to take a very aggressive approach to mitigation stabilizing atmospheric CO_2 at no more than 350 ppm, the real precautionary message is that we ought to be mitigating climate change to the greatest extent possible so as to limit the risk of climate catastrophe(s).

Taking a precautionary approach to climate policy guided by the Catastrophic Precautionary Principle makes it clear that we should not be willing to risk the catastrophic outcomes that are likely to come with even the current level of atmospheric CO_2 and other GHGs. It has the potential to make crystal clear the demandingness of our precautionary obligations to mitigate climate change aggressively. It leaves no room for arguments that keeping global mean surface temperature below 1.5/2°C is sufficient for avoiding dangerous climate change because the risks of this path are simply too high, they are catastrophic. At the same time precautionary mitigation policy will have to spell out how to determine what specific mitigation measures should be implemented, who should take these measures, and who should pay for these measures. It is here that diverse moral and political reasons will have to come into play. Many mitigation strategies can simultaneously promote other desirable ends. For example, providing people with carbon-neutral energy who have previously relied on fires for illumination, cooking, and heating promotes the diverse goals of sustainable development while also mitigation of climate change. Mitigation strategies that have such co-benefits are almost certainly desirable. Cost-effectiveness may even have a role to play in guiding us to choose between equally morally appealing options. For certainly there is no reason to promote expensive mitigation strategies if cheaper strategies are available that also simultaneously satisfy other moral and political aims. From the perspective of the Catastrophic Precautionary Principle, what matters is that a suite of mitigation measures are implemented that together put us on a pathway to avoid catastrophic climate impacts. As Hansen has pointed out, the pathway to 350 ppm is formidable, but with steep emissions reductions and complementary strategies we could get there.[14] However, no matter how successful we are at mitigating climate change, our unfortunate reality is that we have not done enough to eliminate the risk of catastrophic climate impacts via mitigation. Adaptation hence must be a key part of precautionary climate policy.

3 A Precautionary Approach to Adaptation

The gap between where we are and where we should be with our mitigation efforts, the evidence we have of how harmful climate change may become in the meantime, and the harmful climate impacts we are already seeing all suggest we need to find ways to adapt to our changing climate if we are to avoid climate catastrophe. Adaptation involves trying to reduce the harmfulness of whatever climate impacts do in fact occur. So while reducing fossil fuel consumption is a form of mitigation, building sea walls to minimize the harmfulness of sea level rise is a form of adaptation. The extent to which further climate change is mitigated will determine the need for adaptive precautionary measures. There are many different possible adaptation measures in part because climate change impacts are and will be so diverse, and for any given impact there are likely many ways to affect how harmful it is. For example, precautionary measures aimed at preventing sea level rise from being harmful could include the building of sea walls, plans to relocate vulnerable populations, strategies to address the harmful impacts of salt water intrusions into ground water, and so on. Other adaptive precautionary measures include developing drought-resistant crops, implementing evacuation plans for coastal cities vulnerable to hurricanes, developing medicines and ways of distributing these to address the spread of tropical diseases (e.g., malaria), and developing desalinization technologies to provide vulnerable populations with fresh water.

Determining appropriate adaptation strategies and policies as part of a precautionary response to climate change may therefore be much more complex than determining appropriate mitigation policies. It will involve understanding and predicting the causal mechanisms that will lead to potentially harmful climate impacts so that the harmfulness of these impacts can be minimized. It will also involve bringing in a wide range of other morally and politically relevant principles and considerations. Here again the Catastrophic Precautionary Principle is helpful insofar as it can set a clear floor for minimally acceptable adaptation policy. However, to see how and why this is so, it is first important to better understand just what adaptation is and consists in since adaptation can be "a deceptively simple concept,"[15] which is what I do in the following subsection before going on to discuss what an appropriate approach to adaptation as precaution might look like.

3.1 Complexity of Adaptation[16]

Since climate change will have both direct and indirect human impacts, adapting to climate change will involve addressing not only physical climate impacts but also derivative effects of these physical impacts. Climate impacts will also vary greatly across regions not just because of the differential warming and the associated physical changes across the globe but because of variation in system adaptability and adaptive capacity. How well people will be able to adapt to specific climate impacts depends on many factors, including the availability of natural resources, technological and financial capital. How adaptation is understood depends on who or what we value and who or what is adapting.[17] For example, adaptation measures meant to protect coastal property may not allow local ecosystems to adapt to rising sea levels, and adaptation measures meant to protect coastal wetlands may be harmful to local landowners. Further, since people will be differentially affected by climate impacts depending on when and where they live or exist, what constitutes adaptive measures will vary with context.

Some adaptive measures, such as building sea walls, are autonomous, conscious responses to climate change, while others, such as the earlier migration of birds, are automatic. Adaptation also involves both proactive responses, which involve anticipation and planning, and reactive responses.[18] Even inaction may be an implicit or explicit adaptive response. While many adaptive measures are or will be very local, directed towards specific hazards, the process of adaptation can also aim to make society more resilient to a range of influences by increasing adaptive capacity.[19] Another way of increasing adaptive capacity is to focus on building ecological resilience.[20] Adaptation can also be approached in different ways depending on the time scale considered. Adaptation may take the form of a risk-based, short-term approach, a somewhat longer-term social vulnerability approach, or an intergenerational resilience approach.[21]

It is not surprising then that adaptation has been defined in different ways throughout the literature.[22] I follow the common definition where adaptive measures are understood as aiming to reduce the harmfulness of climate impacts.[23] As such, adaptation is inherently linked to precaution, which is also about preventing or minimizing harmful outcomes. Adaptation is hence a counterpoint to mitigation, as it can serve as a category of precautionary response to climate change aimed at minimizing the harmfulness of whatever climate impacts come to pass (i.e., address whatever climate change we failed to prevent via mitigation).

Sometimes compensation is also considered to be a form of adaptation insofar as there may be a need to compensate people who are harmfully affected by climate change,[24] but when understanding adaptation as a form of precaution, compensation should be viewed as a separate category since it aims to compensate for the harmfulness of climate impacts after the fact. Compensation is not a precautionary measure *against* climate change in the way adaptive measures are. (It is also hard to fathom that there will be any meaningful ways *to* compensate for climate catastrophe.) While some adaptive measures are reactive, all adaptive measures are preventative in the sense of aiming to prevent climate impacts from being harmful. For example, building a sea wall is an anticipatory adaptive measure whereas providing fresh water to people affected by salt-water intrusion is a reactive adaptive measure even though it aims to prevent a given climate impact, in this case sea level rise, from being harmful. Providing financial compensation to a family whose child died due to a climate impact, such as lack of access to fresh water following salt-water intrusion, should not be considered as an adaptive measure since this would not prevent the given climate impact from being harmful but instead would serve as compensation for a harmful effect after the fact.

Even if adaptation and compensation are understood to be distinct in this way, some authors argue we ought to approach both adaptation and compensation together. Catriona McKinnon, for example, argues we ought to set up an intergenerational climate change compensation fund now "in order to provide future generations with funds and resources adequate to compensate them for the harm at which we are possibly (with respect to climate catastrophes) and probably (with respect to non-catastrophic but still harmful climate change) putting them at risk."[25] She reaches this conclusion because she is motivated to discern what justice, in this case primarily corrective justice, requires in the case of climate change. So while I agree with much of McKinnon's discussion of precaution in the face of climate change insofar as we both argue precaution demands aggressive mitigation efforts, our views depart in our treatment of adaptation. We both ultimately argue that we ought to be saving so as to fund adaptation efforts in the future, but because I am committed to a strictly precautionary framework I maintain a focus on adaptation (as distinct from compensation) since compensation is not a form of harm prevention (and hence is not a form of precaution).

Part of why adaptation is a deceptively simple concept is that climate impacts are hard to identify as such and it can be hard to distinguish adaptation from development and other issues. First, as was suggested in Chapter 1, particular anthropogenic climate impacts are often difficult to separate from natural climatic events. Second, a simple example elucidates why adaptation is a complex issue that is hard to disentangle from how we think about development and other related issues. Imagine a three-year-old girl living in the Sahel region of Africa in 2080. The region she lives in has been devastated by drought, owing at least in part to climate change. Food and fresh water have become increasingly difficult to acquire. Despite her family's efforts, she suffers from malnutrition. Eventually she dies as a result. The key point is that both adaptation and development policies aim to prevent outcomes such as this one. A program aimed at providing food and clean drinking water to at-risk families like the one in this example could be part of a development program and/or an adaptation program.

Adaptation and development both aim to protect people from harmful external influences. Development and adaptation goals therefore may be hard to distinguish in both theory and in practice. Improving health services, for example, may be a goal of both development and adaptation, as this will improve the lives of people in developing countries while bolstering adaptive capacity (by, for example, reducing the impacts of increased tropical diseases due to warming). It may not even be useful even to think of adaptation policy as independent of other policies because of the interconnections between different kinds of policies.[26] Integrating these policy arenas may help ensure coordinated institutional responses, which in turn may be more effective, but doing so further blurs the definition of adaptation. The flip side of this, however, is that increased focus on adaptation will help vulnerable people no matter what is actually causally contributing to particular harmful effects.[27]

Climate change will have much greater negative direct and indirect impacts in developing nations.[28] These nations will not only bear the greatest physical impact of climate change but are also less capable of adapting to both direct and indirect impacts. This again highlights the connection between development and adaption and is also why some argue that the most vulnerable should be given special attention because of their vulnerability and need for resources.[29] Development may be able to be promoted in ways that simultaneously improve adaptive capacity and vice versa. We might even think of adaptation as "climate-resilient development"[30] and move from thinking about adapting to climate change to thinking about adapting *with* climate change.[31] What unifies "adaptation" as a concept then is that adaption aims to minimize or reduce the harmfulness of climate change. It is in this sense that adaptation is inherently linked to precaution, which is also about preventing or minimizing harmful outcomes.

3.2 Adaptation as Precaution Against Catastrophes

The Catastrophic Precautionary Principle justifies precautionary adaptation measures against climate impacts that would affect many millions of people. On the one hand we may think this only applies to particular climate impacts that threaten to cause severe harm to many millions of people, such as a drought in the Sahel region in a particular year. On the other hand we may think that as soon as the full range of harmful climate impacts threaten to severely affect many millions of people, the Catastrophic Precautionary Principle calls for adaptive measures against all harmful climate impacts. Given that climate change is the driving cause for concern, the latter conclusion is most plausible because what the Catastrophic Precautionary Principle demands is that we try to avert many millions of people from being affected by climate change, though certainly it also justifies taking precautionary measures against specific climate impacts that threaten to be catastrophically harmful on their own. A precautionary approach to adaptation that is driven by the Catastrophic Precautionary Principle should therefore take a wide viewpoint and consider climate change as posing threats of catastrophic impacts as soon as all climate impacts collectively threaten many millions of people. This implies that we ought to be quite aggressive and will have to take a global perspective in approaching adaptation policy since climate change threatens to be harmful in so many ways, in so many places. So while on the one hand the Catastrophic Precautionary Principle may seem too weak a tool to guide adaptation policy, given our current context it turns out to be very powerful. Given the scope of potentially harmful climate impacts, it is very demanding indeed.

To illustrate how the Catastrophic Precautionary Principle could be used as a powerful driver of adaptation policy I will suggest how it might be integrated into the Cancun Adaptation Framework, adopted in December 2010 by many of the world's nations, as this framework continues to drive UNFCCC adaptation policies.[32] As part of the UNFCCC's 16th Conference of Parties, this framework has the objective of "enhancing action on adaptation, including through international cooperation and coherent consideration of matters relating to adaptation under the Convention."[33] The provisions of the framework include, "[p]lanning, prioritizing and implementing adaptation actions," "[s]trengthening institutional capacities and enabling environments for adaptation, including for climate-resilient development and vulnerability reduction," "[b]uilding resilience of socio-economic and ecological systems," and "capacity-building for adaptation, with a view to promoting access to technologies, in particular in developing country Parties."[34] The Cancun Adaptation Framework lays a foundation for cooperation and coordination of adaptation policies and initiatives, while acknowledging the linkages between adaptation and development. One of the initial tenets of agreement that comes out of the 16th Conference of Parties in fact affirms "the legitimate needs of developing country Parties for the achievement of sustained economic growth and the eradication of poverty, so as to be able to deal with climate change."[35]

As it stands, the Cancun Adaptation Framework does not make any reference to precaution. It does, however, include the following clause inviting Parties to undertake:

> Enhancing climate change related disaster risk reduction strategies, taking into consideration the Hyogo Framework for Action, where appropriate, early warning systems, risk assessment and management, and sharing and transfer mechanisms such as insurance, at the local, national, subregional and regional levels, as appropriate.[36]

The Hyogo Framework for Action addresses building the resilience of nations and communities to disasters. The stated expected outcome of the Hyogo Framework is "[t]he substantial reduction of disaster losses, in lives and in the social, economic and environmental assets of communities and countries."[37] The connection between adaptation and the reduction of disaster losses is clear. Anthropogenic climate change affects the otherwise natural climatic system in such ways that there are and will be more frequent environmental (a.k.a., "natural") disasters, among other effects.

The three strategic goals of the Hyogo Framework are also consistent with the general aims of adaptation policy:

a   The more effective integration of disaster risk considerations into sustainable development policies, planning and programming at all levels, with a special emphasis on disaster prevention, mitigation, preparedness and vulnerability reduction.

b   The development and strengthening of institutions, mechanisms and capacities at all levels, in particular at the community level, that can systematically contribute to building resilience to hazards.

c   The systematic incorporation of risk reduction approaches into the design and implementation of emergency preparedness, response and recovery programmes in the reconstruction of affected communities.[38]

This is a place where the Catastrophic Precautionary Principle could be used specifically to strengthen the UNFCCC's adaptation policies. As it stands there are no explicit commitments in the Cancun Adaptation Framework specifying the extent to which adaptation is required or the limits of acceptable losses. Minimally the UNFCCC should be willing to commit to avoiding catastrophic outcomes, which is what including the Catastrophic Precautionary Principle would do.

The Cancun Adaptation Framework's commitment to the Hyogo Framework may provide a path for strengthening the UNFCCC's stance on adaptation. The first strategic goal of the Hyogo Framework puts a "special emphasis on disaster prevention."[39] Since disasters are often coextensive with catastrophes, this goal could be strengthened and reinterpreted as a commitment to the Catastrophic Precautionary Principle. Future UNFCCC adaptation policies could extend their commitment to the Hyogo Framework specifying that the minimal goal is preventing climate impacts from becoming catastrophically harmful, per the definition of catastrophe in the Catastrophic Precautionary Principle. Explicit commitment to the Catastrophic Precautionary Decision-Making Framework would provide a pathway for realizing this commitment.

Stepping back, the Cancun Adaptation and Hyogo Frameworks both emphasize risk reduction, which is closely linked to, though distinct from, precaution. Whereas risk reduction strategies are often understood to involve statistical calculations that require a concrete understanding of the issue in question, a precautionary approach is usually understood as entailing that we sometimes ought to act in advance of scientific certainty to prevent harmful outcomes. Climate change is such a complex phenomenon that we will often not be able to be certain when, where, or how climate impacts will be harmful, though we may have strong reasons to believe implementing anticipatory adaptive precautionary measures will reduce the harmfulness of eventual impacts. For example, we may not know the extent to which fresh water resources in the Sahel or even the southwest United States will become limited in coming years or exactly which sources will be implemented in which ways, yet we may have strong confidence that increasing water storage and transport capacities in this region will have significant net benefits on the whole.

Making explicit the precautionary approach that is currently implicit to the Cancun Adaptation and Hyogo Frameworks in future UNFCCC adaptation policies would allow more room for addressing uncertainty in risk-reduction strategies. This could partially be accomplished by making it explicitly clear that what is meant by the phrase, "enhancing climate change related disaster risk reduction strategies," which is currently in the Cancun Adaptation Framework, is a precautionary approach to risk reduction as well as a commitment to the Catastrophic Precautionary Principle. It would also have to be made clear that this implies an approach to risk management that does not always require certainty. Committing to the Catastrophic Precautionary Principle would provide clear guidelines about when scientific uncertainty may be used as a reason to take precautionary measures in advance of scientific certainty. Namely, the Catastrophic Precautionary Principle specifies that if a threat of harm is catastrophic, the mechanism by which the threat would be realized is well understood and the conditions for the function of the mechanism are accumulating, precautionary measures should be taken in an effort to avoid catastrophe.

We know, for example, that climate change will increase sea level rise, though it is at this point genuinely uncertain how fast sea levels will in fact rise. If, for example, we also have good reasons to think that many millions of people in Bangladesh would be seriously harmfully affected by 2 meters of sea level rise and we suspect, though we are not certain, that sea levels will rise by at least 2 meters at some point in the next century, we are warranted in implementing precautionary measures in Bangladesh against impending sea level rise. Exactly what precautionary measures we ought to implement will depend on a lot of factors and may at first involve merely learning more about how sea level rise would affect Bangladesh and what measures could be taken to protect people from being severely harmfully affected. The Catastrophic Precautionary Decision-Making Framework can at least guide decision makers as they navigate the complex task of determining an appropriate course of precautionary action. While a focus on risk reduction and/or resilience building might also support precautionary action in this kind of case, neither of these concepts explicitly makes room for erring on the side of caution in the face of scientific uncertainty. One of the reasons precaution is such a pervasive concept in the environmental arena is that it steers us away from an over reliance on cost-benefit analysis and decision-making procedures that implicitly require that we know what will happen before we act. Together a precautionary approach, the Catastrophic Precautionary Principle, and the Catastrophic Precautionary Decision-Making Framework capture the essence of what the Cancun Adaptation Framework already says in a way that has the potential to be both more accommodating of uncertainty and much more concrete and aspirational in its goals by establishing a clear aim for mitigation policy: preventing climate impacts from severely harming many millions of people. The Catastrophic

Precautionary Principle hence may be leveraged in both guiding UNFCCC policies and more specific policies like the Cancun Adaptation Framework to set clear bounds on acceptable climate risk in a way that accommodates the uncertainty that is inherent to our understanding of climate change.

4 Geoengineering as Precaution?

A third category of response to climate change tries directly to impact the climate system. Whereas mitigation efforts may be seen as trying to prevent further climate change and adaptation measures try to address whatever climatic changes occur, this third category of responses operates in the space between these two ends, trying to reduce the influence of increased GHGs in the atmosphere. This category of response is often referred to as geoengineering because it involves intervening in the global climatic system itself, rather than the drivers or impacts of climate change. Geoengineering strategies are commonly broken down into two categories: CO_2 removal and solar radiation management. CO_2 removal is exactly what it sounds like. Methods of removing CO_2 from the atmosphere range from reforestation to ocean fertilization to point-source carbon capture and storage. Solar radiation management is also just what it sounds like. It aims to manage the amount of incoming solar radiation entering the atmosphere, the idea being that reducing the amount of energy entering the climate system could reduce global temperatures. Methods of solar radiation management range from painting roofs white to increasing cloud cover over the oceans to stratospheric sulfate injection. While mitigation and adaptation should clearly play a role in a precautionary response to climate change, it is less clear what role, if any, geoengineering should play. I hence explore below at length whether geoengineering strategies should be considered as appropriate precautionary measures against climate change. Here again we will see that other moral and political considerations should play a key role in determining the place of geoengineering strategies, but the Catastrophic Precautionary Principle places strict limits on how and when such strategies may even be brought to the table.

Both mitigation and adaptation should play a central role in a precautionary approach to climate change, but what about geoengineering? A naïve interpretation may at first glance seem either to support geoengineering as possible precautionary measures against climate change or reject such measures as being too risky.[40] Kevin Elliott rightly argues that because there are so many different geoengineering strategies and so many formulations of the precautionary principle, there is no straightforward relationship between the two.[41] While the first concern – the existence of many different geoengineering strategies – makes understanding the role of geoengineering in a precautionary response to climate change challenging, the Catastrophic Precautionary Principle avoids the second worry. In fact, the Catastrophic Precautionary Principle gives us clear guidance for determining if a particular geoengineering strategy may be a part of an appropriate precautionary response to climate change insofar as it requires that precautionary measures do not themselves introduce or exacerbate threats of catastrophe. In this way it may help us identify one of the limits on morally acceptable geoengineering strategies, that is what strategies are "sufficiently safe" to be considered for implementation along the lines suggested by the Royal Society.[42] As in the last section, I start by exploring the concept of geoengineering in the first subsection before going on to examine the precautionary limits on geoengineering. A final subsection then discusses a lingering question about last resorts.

4.1 Geoengineering as a Category of Response to Climate Change

In order to see how the Catastrophic Precautionary Principle can help us assess an appropriate role for geoengineering in climate policy, we have to understand more clearly the spectrum of strategies that fall under this broad heading – wrestle with the meaning of "geoengineering" – and work through the principle's implications for geoengineering as a category of response and for particular geoengineering strategies. Geoengineering is a complex concept about which there have arisen many ongoing debates.[43] In fact, the term has been used for decades or even centuries to apply to a wide range of things.[44] Currently a commonly accepted definition of geoengineering in the context of climate change is that it is the intentional manipulation of the global climatic system either through carbon sequestration or solar radiation management, hence in the context of climate change "geoengineering" may more aptly be described as "climate engineering." In either case, a huge range of strategies or techniques falls under these general banners.[45]

There are both biotic and abiotic carbon sequestration strategies. Biotic carbon sequestration strategies aim to take advantage of the photosynthetic activity of living organisms to store carbon in plants and soils. While some of these strategies may more aptly fall into the category of mitigation, when undertaken with the aim of sequestering carbon rather than slowing emissions such strategies arguably fall into the category of geoengineering. Four examples of biotic carbon sequestration are promoting forest growth, soil management for carbon storage through changing agricultural practices, burying biomass – primarily biochar (i.e., charcoal) – in soils, and ocean fertilization, which promotes algal growth and deep-sea sequestration of biomass.[46]

Abiotic methods of carbon sequestration, on the other hand, employ non-living chemical reactions to sequester CO_2 from the atmosphere. Strategies include managed carbon capture and storage, and acceleration of natural weathering, which converts CO_2 into stable carbonates by reaction with metal oxides, though the overall cost and energy requirements make accelerated weathering unfeasible with current technologies.[47] Point source carbon capture and storage aims to scrub CO_2 emitted from industrial coal and gas combustion, concentrating and pumping it into geological formations on land or in the ocean. While there is the potential for up to 545 GtC to be stored this way over time,[48] there are serious risks associated with the possible leakage of stored carbon. Strategies are also emerging that would capture CO_2 from ambient air using a sorbent material, including via so-called "artificial trees." This approach aims to capture CO_2 released from diffuse sources such as automobiles and deforestation, though such methods are currently energy and cost prohibitive compared to traditional mitigation efforts.[49]

Solar radiation management strategies are very different from carbon sequestration strategies; they aim to reduce solar energetic input to the Earth's atmosphere and surface, where it turns to radiant heat. Proposed methods include outer-atmosphere reflectors, marine cloud brightening, stratospheric aerosols, and whitening surfaces in cities, oceans, and deserts.[50] The motivation for outer-atmosphere reflectors or putting sunshades in space is that incoming solar radiation would only have to be reduced by 1.8% to offset a doubling of pre-industrial atmospheric CO_2 concentrations. The remaining solar radiation strategies all aim to increase the Earth's albedo so as to reflect solar radiation either from the stratosphere, lower atmosphere, or Earth's surface. Stratospheric aerosols, or stratospheric sulfate injection as it is also sometimes called, aims to mimic the effects of large volcanic eruptions, which have been shown to cool the planet temporarily. Several different delivery methods have been proposed including balloons and high-level aircraft. Marine cloud brightening, or enhanced cloud albedo, aims to increase the reflectivity of low-level marine stratiform clouds through either mechanical (e.g., sea sprayers) or biological (e.g., side effects of phytoplankton growth) mechanisms. Finally, whitening a wide range of terrestrial surfaces from roofs to grasslands are possible strategies of surface albedo enhancement. Nevertheless, the climatic effects of all of these methods remain highly uncertain and difficult to test before widespread adoption.[51]

Clearly there is a huge range of strategies that have been categorized as forms of geoengineering, but does "geoengineering" capture a morally or practically relevant set of responses to climate change? Dale Jamieson argues that we should not only maintain the distinction between carbon sequestration and solar radiation management but should also re-categorize these strategies because they target different causal processes within the climatic system.[52] Jamieson even seems to suggest that we may have reasons to abandon "geoengineering" as an umbrella term. Carbon sequestration should, he argues, be re-categorized alongside mitigation as "abatement," since the aim of both traditional mitigation and carbon sequestration is to abate increased atmospheric concentrations of GHGs. He argues that "abatement" is the term that we usually associate with reducing the emissions of pollutants, whereas "mitigate" usually refers to moderating the severity of the effects of pollution. "Abatement" should therefore become the term of choice for strategies aimed at minimizing the extent of climate change through reductions in atmospheric GHG concentrations. Jamieson argues solar radiation management, on the other hand, is more aptly described at a form of mitigation than is the reduction of GHG emissions since it aims to mitigate the effects of increased atmospheric concentrations of GHGs, so-called "carbon pollution."[53] "Geoengineering" in Jamieson's view is thus not a helpful term since it obscures the distinction between abatement (a.k.a., mitigation) and mitigation (e.g., solar radiation management).

While I agree with Jamieson that what we mean when we talk about mitigating climate change does not track the way we otherwise talk about mitigating pollution-driven environmental problems, I think it is simply too late for us to change course and adopt a new way of talking about possible responses to climate change. As a term, "mitigation" is simply too deeply entrenched in our thinking about climate change. Furthermore, geoengineering does capture a practically and morally relevant category of possible responses to climate change. Jamieson worries that as a category geoengineering is supposed to capture strategies that have a global reach but at least some geoengineering strategies can be implemented at such a small scope that this simply is not the case.[54] It is certainly true that planting or saving one tree will not sequester enough CO_2 to meaningfully affect the greenhouse effect, and, similarly, painting one roof white will not reflect enough solar radiation to reduce global temperatures even a little bit. However, this does not undermine the fact that at some scale all of the above-discussed geoengineering strategies could, at least in theory, affect the global climatic system; this is why they have all come to be associated with the term "geoengineering" in the first place. Just as not all mitigation efforts actually mitigate (or abate, to use Jamieson's term) climate change – think for example of an individual choosing to cycle rather than drive to work on any given day in the name of mitigation[55] – the fact that not all geoengineering efforts actually affect the global climate system does not undermine the term. Effective mitigation policies and strategies will actually mitigate (or abate) climate change, and effective geoengineering policies and strategies will actually have a discernable effect on the global climatic system.

I am similarly skeptical of claims that solar radiation management is importantly different from carbon sequestration because of issues of scale. It seems to be a common assumption that, in general, solar radiation management will usually (or even always) have a greater impact on global climate than carbon sequestration. I think the idea is that planting trees simply cannot compare to stratospheric sulfate injection or marine cloud brightening. While the fact that painting roofs white would be a form of solar radiation management if widely implemented may be enough to counter this notion, recent research illustrates that the scope of effects of reforestation efforts could have very significant effects on par with those of solar radiation management. Abby Swann and her colleagues, for example, have shown through modeling that large-scale afforestation in northern mid-latitudes, while having a minimal effect on global temperatures, could have significant climate impacts particularly with respect to precipitation changes because of its effects on global circulation patterns.[56] This example not only illustrates the potential of carbon sequestration strategies to have significant (and potentially harmful) global climatic effects, but also illustrates why we should not be too quick to assume solar radiation management is inherently riskier than carbon sequestration. It may be tempting to think that there cannot be anything bad or harmful about planting a few trees, but if enough trees are planted we could disrupt important climatic systems with unknown consequences. We should not forget that what got us into the situation we are in now were and are seemingly innocuous activities such as cutting down forests and burning fossil fuels to power automobiles. The only difference between these GHG-emitting activities and geoengineering strategies is that we are not now, and historically people were not, *intentionally* aiming to modify the global climate with these activities.[57]

Geoengineering should be maintained as a practically and morally relevant category of possible responses to climate change because it captures a class of strategies that aim to intentionally affect the global climatic system. Whereas mitigation efforts aim to reduce further contributions to climate change and adaptation strategies aim to reduce the harmfulness of eventual climate impacts, geoengineering aims to tinker with the climatic system itself by affecting causal mechanisms in this system. Though some mitigation strategies may be importantly similar to carbon sequestration strategies, this only underscores the need for us to assess each possible strategy for addressing climate change individually. It may turn out that some carbon sequestration strategies are in fact less risky than some mitigation strategies. We must also remember that all responses to climate change are just that. There will likely be some similarity between different kinds of responses simply because they all aim to minimize the extent and/or harmfulness of climate change. What this section has established, however, is that "geoengineering" as a term captures, in at least some ways, a distinct set of responses to climate change. While ultimately each geoengineering strategy will have to be assessed on its own in terms of whether or not it could ever be an appropriate precautionary measure against climate change, I can now offer some initial thoughts about the particular worries geoengineering strategies will come up against, as seen through the Catastrophic Precautionary Decision-Making Framework.

4.2 Precautionary Limits on Geoengineering

The key element of the Catastrophic Precautionary Principle to consider in order to determine if a given geoengineering strategy may be an appropriate precautionary measure against climate change is its requirement that appropriate precautionary measures must not create further threats of catastrophe and must aim to prevent the potential catastrophe in question. It is pro tanto wrong according to the Catastrophic Precautionary Principle knowingly to act in a way that poses a threat or threats of catastrophe. This does not mean that circumstances will never be such that we cannot take significant risks, but if we do engage in risky activities we usually must also implement precautionary measures to mitigate the possibility of catastrophic outcomes. At the same time, unless other reasons come into play, it is pro tanto wrong knowingly to introduce, reinforce, or further a threat of catastrophe. This means that unless there is an overriding reason, ultimately we should not act in ways that pose threats of catastrophe.

Precautionary measures taken against a particular threat of catastrophe are not exempt from the Catastrophic Precautionary Principle merely because they are precautionary measures. Taking precautionary measures involves acting, so the Catastrophic Precautionary Principle says that pro tanto precautionary measures should not introduce (unmitigated) threats of catastrophe. The Catastrophic Precautionary Principle and the Catastrophic Precautionary Decision-Making Framework imply that when we are faced with a particular threat of catastrophe and we are assessing what an appropriate course of precautionary action is, our focus should be not only on eliminating or reducing the given threat of catastrophe but also on averting catastrophic outcomes in general. In order not to violate the Catastrophic Precautionary Principle precautionary measures themselves pro tanto should not pose new or further threats of catastrophe. Therefore if there are available precautionary measures that would eliminate or sufficiently mitigate the given threat without creating or furthering threats of catastrophe, the Catastrophic Precautionary Principle will direct us to choose among these options.

The relevant worry about all geoengineering strategies is that they involve manipulating global systems in a way that could have potentially catastrophic side effects. In order to violate the Catastrophic Precautionary Principle the side effects of a geoengineering strategy would have to severely threaten many millions of people, which, given the complexities and scale of the systems involved, seems plausible for many geoengineering strategies. Put one way, this means that some geoengineering strategies should be considered as possible precautionary measures against climate catastrophes if these strategies do not present any of their own or exacerbate existing threats of catastrophe. On the other hand, the burden of proof of safety is set extremely high by the Catastrophic Precautionary Principle because it implies that a precautionary measure cannot introduce a new threat of catastrophe however remote or unlikely that threat is. In fact, it is plausible that the implementation of any geoengineering strategy that would have a discernible influence on the global climate would likely threaten to have potentially catastrophic side effects. The Catastrophic Precautionary Principle, remember, states that it is enough to understand the mechanism by which the threat of harm would be realized, that the conditions for the function of the mechanism are accumulating, and that likelihood not be taken into account (at least not at the stage of determining whether a threat should be deemed catastrophic). This implies that precautionary measures must meet these standards and cannot introduce even remote or unlikely threats of catastrophe if they are to fully satisfy the Catastrophic Precautionary Principle. In this case the mechanism would be a potential side effect of a geoengineering strategy and the implementation of the strategy would constitute the realization of the conditions for the function of the mechanism. So if we have any reasons to suspect a form of geoengineering will have potentially catastrophic side effects, it will violate the Catastrophic Precautionary Principle and cannot constitute an appropriate precautionary measure against climate change.

One may object to this line of argument by claiming that in many cases it would be reasonable to take a precautionary measure that raises a threat of catastrophe if the magnitude or likeliness of the "new" threat of harm is less than the threat being addressed. The Catastrophic Precautionary Principle, however, does not condone precautionary measures that themselves raise threats of catastrophe because it is justified on the basis that such threats are pro tanto morally unacceptable. While some authors have interpreted the precautionary principle as guiding us to reduce relative threats, I reject such interpretations because they ultimately fail to provide clear and consistent moral guidance. If all a precautionary principle required of us was that we take precautionary measures that reduce a threat of harm or even a threat of catastrophe, we might be justified in taking superficial and ineffective measures that we know to raise their own threats of catastrophe when there are other precautionary measures available that do not invite such threats. We may satisfy such a precautionary principle, for example, by implementing a geoengineering strategy that we know poses new threats of catastrophe and abandon mitigation efforts, but if we believe we have moral reasons to avoid the threats of catastrophe posed by climate change, we also have reasons to avoid the threats of catastrophe posed by any risky geoengineering strategies. Trading one set of risks for another does not satisfy our obligation to reduce the first risks because it involves doing what we accept to be wrong. Addressing climate change with space mirrors before we are sure doing so does not pose any of its own threats of catastrophe for future generations, for example, would be like justifying playing Russian roulette with a stranger rather than our own child. While we may have extra reasons not to risk killing our own child or the current generation, from the perspective of morality it does not matter *when* a catastrophe threatens to ensue for us to have strong reasons to avoid engaging in what we know to be a catastrophically risky activity.

Implementing a geoengineering strategy in a way that would have a discernable effect on climate change would involve, by definition, manipulating the global climate. It is hard, though not impossible, to imagine that any negative side effects of such a wide-scale implementation would not pose any threats of catastrophe. Painting a few roofs white may not be catastrophic, but before a massive roof-painting campaign is initiated the bar will be set very high for proving that the solar radiation reflected by white roofs will not have catastrophic consequences. Before we inject sulfur into the stratosphere the Catastrophic Precautionary Principle requires that we have strong evidence that deployment of this geoengineering technology would not create new or exacerbate existing threats of catastrophe. The Catastrophic Precautionary Principle does not rule out geoengineering strategies as a possible precautionary measure against the catastrophes climate change threatens, but it does require we prove a geoengineering technique to be safe insofar as it will not introduce new threats of catastrophe, before implementation. This is a very high bar for measures meant intentionally to manipulate the climate system on a global scale.

From this perspective geoengineering strategies that can be tested and implemented at smaller scales are more likely to be acceptable than those that can only be implemented at a wider scale. Forms of carbon sequestration such as biochar, for example, may not pose catastrophic risks at small scales. In fact, several colleagues and I assessing climate engineering strategies from diverse perspectives agree that biochar is amongst the least risky carbon sequestration strategy both because it has only local immediate effects and does not appear to pose significant long-term risks.[58] The problem, however, is that one of the reasons biochar looks less risky is because of its limited scale and hence limited ability to sequester significant amounts of carbon. A close evaluation of biochar hence may reveal that it passes the bar set by the Catastrophic Precautionary Principle and hence may be implemented as *a* precautionary measure against climate catastrophe, but because of its limited scale it will only be able to play a small part in a comprehensive precautionary strategy.

However, scale is not the only relevant factor in determining whether a geoengineering strategy is sufficiently safe. Ocean fertilization techniques, for example, may be able to be implemented at small scales, but evidence to date suggests not only that such techniques are unlikely to be effective at sequestering carbon over long time periods but also that they create significant risks to ocean ecosystems, the ramifications of which we do not sufficiently understand.[59] Strategies that involve the geological burial of liquid CO_2 can also be implemented at small scales, though these strategies raise significant questions about long-term risks as sudden releases of stored CO_2 (e.g., as could be caused by an earthquake or other geological event) could have catastrophic ramifications, hence the burden of proof of safety of storage facilities/locations would be very high.[60]

We have to be very careful not to overlook the fact that even seemingly small-scale carbon sequestration projects are meant to have an effect on the global climate. It is this very fact that raises a red flag from a precautionary perspective. If climate change has taught us anything it is that seemingly inconsequential actions can add up to have devastating, potentially catastrophic implications collectively. Climate policies governing geoengineering hence have to look at both the risks of any particular carbon sequestration project and the risks posed by all carbon sequestration projects collectively. Given how far we are from implementing effective mitigation policies and reining-in our GHG emissions, it is reasonable that we are exploring ways to pull carbon back out of the atmosphere to minimize the extent of eventual climate change. The Catastrophic Precautionary Principle can help us see just how high the standard of evidence should be before we let ourselves feel satisfied that any particular carbon sequestration project and our collective sequestration strategy is sufficiently safe.

This is even truer when we turn to solar radiation management, in part because of what the Royal Society calls the termination problem.[61] In order to maintain lower temperatures solar radiation management must be maintained and implemented in a way that creates the desired balance of solar radiation. Suddenly terminating the implementation of a solar radiation management strategy such as stratospheric sulfate injection could have catastrophic consequences because of the sensitivity of the climatic system to large, rapid changes in incoming solar radiation. In order to prove that a solar radiation management technique does not pose a threat of catastrophe, confidence in a potentially intergenerational implementation scheme would have to be demonstrated, but it is easy to see why such guarantees will be hard to come by. Threats of catastrophe posed by the termination of solar radiation management qualify as threats posed by the implementation of a form of solar radiation management if there is any chance of a termination scenario in which such threats would arise. Given uncertainties about the political stability of geoengineering governance schemes if there is any chance of the termination of a form of solar radiation management (or the abrupt release of massive amounts of CO_2 from a carbon sequestration project) will pose a threat of catastrophe even if the solar radiation management implementation itself carried no threats of catastrophe such a scheme would violate the Catastrophic Precautionary Principle and hence be unacceptable as a precautionary measure against climate change.

In fact, much of the attention around geoengineering has focused on the challenges of governing geoengineering such as that posed by the termination problem. Whatever the status of various geoengineering strategies is, the mere possibility of geoengineering raises questions about governance. Who will decide what the global average temperature should be? Who will regulate geoengineering deployments? How should geoengineering research be governed? The Catastrophic Precautionary Principle and the Catastrophic Precautionary Decision-Making Framework may prove to provide a framework for addressing at least some of the relevant governance issues, though several critical issues are left open for debate. The Catastrophic Precautionary Principle implies, for example, that the burden of proof of safety of particular forms of geoengineering strategies will rest with those who want to implement them. It does so because it is formulated primarily to address existing threats of catastrophe, thereby applying to any precautionary actions taken in response to an existing threat. In this case the Catastrophic Precautionary Principle applies directly to climate change and indirectly to geoengineering as a potential category of precautionary measures against climate change. The Catastrophic Precautionary Principle and the Catastrophic Precautionary Decision-Making Framework imply that whoever bears responsibility for taking precautionary measures against climate change should ensure that whatever precautionary measures are taken meet the Catastrophic Precautionary Principle's criteria and do not create further threats of catastrophe.[62]

All of this said, the Catastrophic Precautionary Decision-Making Framework does guide decision makers to consider all available precautionary measures, which may sometimes include encouraging research into unproven or untested measures such as geoengineering strategies. As the possibility of adequately mitigating climate change fades, geoengineering strategies arguably start to come to the forefront as available precautionary measures. So it seems precaution does warrant research into geoengineering measures as options become limited, so long as such research aims at determining whether such measures satisfy the Catastrophic Precautionary Principle (i.e., do not pose further threats of catastrophe). At this point we simply do not know enough about most forms of geoengineering to be able to make definitive claims about safety. Research into geoengineering therefore is appropriate so long as the limitations imposed by the Catastrophic Precautionary Principle are kept in mind.

Just as the burden of proof of safety is very high for implementing a geoengineering strategy, a similar burden exists for geoengineering research. Here the differences between various forms of geoengineering and the differences between different methods and phases of research become important. Computer modeling of geoengineering strategies does not directly pose any threats of catastrophe (though we may well want to think about the societal implications of such research). Field research of geoengineering strategies, however, may or may not violate the Catastrophic Precautionary Principle. It may be possible, for example, to implement a small-scale test of the impact of painting roofs, to test a small number of "artificial trees" that remove only a small amount of CO_2 from the air, or to have a small-scale test of a marine cloud-brightening strategy in a way that does not pose any threats of catastrophe and may give us valuable data to help us better understand climate change. It may not be possible, however, to field test stratospheric sulfate injection safely because of the global dispersion of even a small-scale deployment of this technology and global risks this brings. In any case, a strong argument would need to be made prior to implementation or testing at any scale in order to satisfy the Catastrophic Precautionary Principle.

Researching geoengineering is therefore consistent with the Catastrophic Precautionary Decision-Making Framework's mandate to explore all possible precautionary measures, but such research must also comply with the Catastrophic Precautionary Principle. As soon as a form of geoengineering is shown inescapably to pose threats of catastrophe it must be taken off the table as a viable precautionary measure against climate change. If we are to follow the Catastrophic Precautionary Principle and take the precautionary course of action, after all, we should act in accordance with precaution and not take new risks with our planet that could threaten both ourselves and future generations in new ways.

4.3 A Lingering Question about Last Resorts

We nonetheless may be left asking ourselves, what if a climate catastrophe is imminently upon us and our only hope of averting it is to inject sulfates into the stratosphere, to brighten clouds in the troposphere, to launch space-based reflective surfaces, to sequester large amounts of CO_2 out of the atmosphere, or implement some other form of geoengineering even though we believe these strategies pose additional threats of catastrophe? What if, in other words, we fail to avert catastrophe using appropriate means and it appears that our only option is radical geoengineering of our planet, reducing the threats of catastrophe that climate change poses in lieu of the threats geoengineering poses? Answering these questions is extremely tricky in part because we are failing to implement precautionary measures that could significantly reduce the risk of climate catastrophe. Yet I have claimed several times that it is hard to imagine the melting of the West Antarctic Ice Sheet *not* being catastrophically harmful, no matter how aggressive we are with mitigation and adaptation. While it might be tempting to think this implies geoengineering should now be on the table as a viable precautionary response to climate change, things are not quite so simple. The Catastrophic Precautionary Principle, remember, does not always warrant taking the most extreme course of precautionary action.

At this point we do have some evidence that the conditions for the functioning of the mechanism that could lead to the inevitable melting of the West Antarctic Ice Sheet are accumulating, hence the Catastrophic Precautionary Principle gives us pro tanto reasons to take precautionary measures against this threat. However, this does not mean that it justifies geoengineering. It might be quite reasonable (after full investigation via the Catastrophic Precautionary Decision-Making Framework) to use this evidence to justify a very aggressive approach to mitigation (as argued above), an aggressive response to adaptation (e.g., that involves advanced planning for minimizing the severe harm caused by sea level rise), and investing significant resources to further studying the West Antarctic Ice Sheet (including whether thresholds for melting have in fact been crossed – remember at this point there is merely some evidence this is the case, hence this is a real risk but we do not yet fully understand all of the causal mechanisms at work). We should face head-on the risks posed by the possible melting of the West Antarctic Ice Sheet, but we should not let our fears about this risk of catastrophe cause us to lose sight of the full range of precautionary measures available to us.[63]

However, should we arrive at a point where safer precautionary measures are proving to be ineffective and a potentially risky geoengineering technique is our only option for averting a climate catastrophe or minimizing the extent of a catastrophic outcome, we are going to have to ask ourselves some hard questions and look beyond the Catastrophic Precautionary Principle. As a pro tanto moral principle the Catastrophic Precautionary Principle cannot and should not be our sole guide in any and all situations, even those involving threats of catastrophe. This is because sometimes there will be other issues or factors of greater moral importance. The Catastrophic Precautionary Decision-Making Framework incorporates an acknowledgement of this in its requirement that we take a wide perspective and take into account other normative considerations. A scenario in which all courses of action involve threats of catastrophe may provide the kind of overriding reason that will make it permissible to take an action that poses a threat of catastrophe, but in such a scenario there is likely to be so much at stake such that the Catastrophic Precautionary Principle alone will not be able to guide decision making.

Elliott suggests that a scenario in which a precautionary principle cannot endorse any precautionary measures should be understood as being self-defeating.[64] His discussion suggests that self-defeating scenarios may enable us to reflect on the strengths and weaknesses of the precautionary principle and push us to consider other moral principles. I think this is exactly right, but I do not see such self-defeating scenarios as problematic for the Catastrophic Precautionary Principle. Rather, they reveal past moral failings and/or tragic circumstances. If future people are forced to endure climate catastrophes, their tragic circumstances will be the result of our moral failings; as such, their moral circumstances will be very different from our own. Catriona McKinnon goes so far as to argue that triaging our failures is not itself a form of justice, for the conditions of justice will not be present in such circumstances.[65] Our focus for now should be on following the Catastrophic Precautionary Principle in an effort to avoid ever reaching the tragic circumstances of a climate catastrophe, of having to triage climate change when justice may no longer be possible.

Nonetheless, this possibility may lead us to ask: should the possibility of inevitable catastrophes on all sides change how we think about geoengineering? Should we "arm the future" with a better understanding of geoengineering strategies by doing more research now?[66] The precautionary stance I am defending answers, "no." What the Catastrophic Precautionary Principle requires of us is to take precautionary measures to avoid being in such a position in the first place. While time may very well be running out, we could avert or at least dramatically reduce the threats of catastrophe posed by climate change through standard mitigation efforts and by preparing to implement adaptation measures in areas most at risk of being catastrophically impacted. The Catastrophic Precautionary Principle does not call for preparing for the worst or "arming the future." Thus precaution, at least as is mandated by the Catastrophic Precautionary Principle, does not at this point justify implementing or even researching geoengineering strategies once these are shown unavoidably to pose their own threats of catastrophe.

What the Catastrophic Precautionary Principle does justify is taking action now to prevent future catastrophes. If painting roofs white, implementing safe carbon capture and storage techniques, or installing reflectors in the desert can safely help us accomplish this goal, these measures should be considered as viable options. The burden lies with those who are responsible for implementing precautionary measures against climate change to demonstrate their safety. However, first and foremost we need to get serious about taking precautionary measures against climate change that we know are safe and would be effective if widely implemented. Namely, we ought to take aggressive action to mitigate climate change and begin implementing adaptive strategies to prevent climate impacts from being catastrophically harmful.

5 Conclusion

Despite the fact that intergenerational risks abound in today's increasingly globalized world, we – as a society – do not yet grasp the complexity of intergenerational risks or how we should address them. Many people share the intuition that we ought to take a precautionary approach to addressing threats of harm, yet it should by now be clear that there has been no consensus about what this means. In a climate of risk we need more carefully to sort out what kinds of risks we ought to accept and live with and what kinds of risks we ought to address via precautionary action. This book has taken us a step in this direction. I hope it has also given us reasons to see that we not only live in a climate of risk, but we live in a climate of hope. Better understanding the risks we face and developing tools for addressing this climate of risk gives us reason to hope all is not yet lost.

There is no one precautionary principle. It is unhelpful to talk or act as if there is. Precaution is a useful concept but only when it is carefully limited. We need to move from talking about "the precautionary principle" to being clear what we mean when we say we are taking a precautionary approach in a given context and to identify and defend specific precautionary principles and precautionary decision-making frameworks that can and should guide us. This will give meaning and force to the concept of precaution in a way that may help us cope with the climate of risk in which we live.

We have especially strong reasons to take precautionary measures against threats of catastrophe, whenever catastrophe threatens to occur. The Catastrophic Precautionary Principle captures this pro tanto moral obligation in a way that is clear and action guiding. Recognizing that uncertainties affect our understanding of and ability to address threats of catastrophe illuminates the need for and helpfulness of the Catastrophic Precautionary Decision-Making Framework. The Catastrophic Precautionary Principle is not (and should not be) the only moral principle that guides decision making even in a case like climate change, but it is relevant whenever we suspect or recognize a threat of catastrophe.

Climate change is one of the most obvious targets of the Catastrophic Precautionary Principle, but this is in part the point of this book. My aim has been to reclaim precaution as a meaningful concept and to illustrate that the Catastrophic Precautionary Principle is *a* precautionary principle that can and should play a powerful role in climate change policy. There are, of course, other issues that might pose threats of catastrophe and about which the Catastrophic Precautionary Principle and the Catastrophic Precautionary Decision-Making Framework can provide guidance – for example, the release of untested chemicals into the environment, or nuclear proliferation, or the possibility of a global flu pandemic all might pose threats of catastrophe. The Catastrophic Precautionary Principle requires us to take precautionary measures against all threats of catastrophe and the Catastrophic Precautionary Decision-Making Framework can help us sort out which threats of catastrophe warrant precautionary action and what this entails on a case-by-case basis.

The main thesis of this book is that we ought to take precautionary measures against threats of catastrophe, and since climate change presents many such threats, we ought to take precautionary measures against potentially catastrophic climate impacts. Given that we already face the threat of climate catastrophe at current atmospheric GHG concentrations, we have very strong pro tanto moral reasons to take significant precautionary measures to mitigate further climate change to the greatest extent possible and to develop and put in place strategies for adapting to those potentially harmful climate change impacts that cannot be avoided. We should also carefully evaluate whether any geoengineering strategies could be implemented without creating or exacerbating threats of catastrophe, in which case they may be implemented as part of the suite of precautionary strategies for minimizing the threat of climate catastrophe. There should be no more excuses or complacency about taking precautionary measures against climate change. Catriona McKinnon sums up the alternative perfectly when she says, "[i]f we fail to make the cuts required within the next couple of decades it is no exaggeration to say that we, the particular cohort of people alive now, will have failed to do justice to the whole human race. For shame."[67] We must now determine not whether we should act, but how we should act so as to fulfill our obligation to prevent the foreseeable catastrophes climate change threatens.

The Catastrophic Precautionary Principle, in its simplicity, is a powerful tool for global climate policy. The UNFCCC is a meaningful and politically ambitious global climate policy framework, but the UNFCCC must make its precautionary goals more explicit if it is to guide us towards preventing climate catastrophe. Embracing an explicit precautionary approach and committing to the Catastrophic Precautionary Principle could and should move the global community to take even more aggressive precautionary measures against climate change. In so doing, climate policy may link with development and other worthy policy goals aimed at preventing catastrophic climate damages.

We have so far failed to take sufficient precautionary measures against climate change, but we must hope it is not too late to avert climate catastrophe. I hope for my son and for the billions who will live after him that it is not too late for us – all of us now and into the future – to take our precautionary obligations seriously. We are living in a climate of risk, but we are also living in a climate of hope.

Notes

1 Hamilton 2013; Vaughan and Lenton 2011; Crutzen 2006; Keith 2000; Shepherd et al. 2009.
2 UNFCCC 1992.
3 Stern 2015; Wagner and Weitzman 2015; Nordhaus 2013.
4 United Nations 2015.
5 Mitchell et al. 2016.
6 Gardiner 2011.
7 Lenton 2011.
8 Wayne 2013; Vuuren et al. 2011.
9 Hansen et al. 2008; Rockström et al. 2009.
10 NOAA 2015.
11 Rockström et al. 2009.
12 Hansen et al. 2015.
13 Hansen et al. 2013.
14 Hansen et al. 2008; Hansen et al. 2015; Hansen et al. 2013.
15 Pelling 2011, 20.
16 This section is adapted from my discussion in the section, "Why adaptation? Understanding climate impacts," from Hartzell-Nichols 2011.
17 Adger et al. 2010.
18 Field et al. 2014; Jamieson 2005; Paavola and Adger 2006.
19 Nelson 2010.
20 Peterson 2010.
21 Eakin et al. 2010.
22 For a discussion of the history of and potential limitations to the term "adaptation," see Orlove 2001.
23 Some people include ways of attaining benefits from climate impacts as adaptive measures (Barnett 2010), though I follow the more standard interpretation that identifies adaptation with reducing the harmfulness of climate impacts.

24 Gardiner 2011.
25 McKinnon 2012, 73.
26 Dovers and Hezri 2010.
27 Jamieson 2005.
28 Stocker et al. 2013; Field et al. 2014; Barros et al. 2014.
29 Dow et al. 2006.
30 Fankhauser 2010, 28.
31 Pelling 2011.
32 Cancun Adaptation Framework 2010.
33 Cancun Adaptation Framework 2010, 4.
34 Cancun Adaptation Framework 2010, 4–5.
35 Cancun Adaptation Framework 2010, 2.
36 Cancun Adaptation Framework 2010, 5.
37 UNISDR 2005, 5.
38 UNISDR 2005, 3–4.
39 UNISDR 2005, 3.
40 Bodansky 1996.
41 Elliott 2010.
42 Shepherd et al. 2009.
43 Vaughan and Lenton 2011; Fleming 2010; Shepherd et al. 2009; Schneider 1996; National Academy of Sciences 1992.
44 Fleming 2010.
45 Cusack et al. 2014.
46 While theoretically promising, field studies have found that very little carbon sinks to the deep ocean for storage (Landry et al. 2000).
47 Anderson and Newell 2004.
48 Pacala and Socolow 2004.
49 House et al. 2011.
50 Vaughan and Lenton 2011.
51 Shepherd et al. 2009.
52 Jamieson 2014, 2013.
53 Jamieson 2014.
54 Jamieson 2013.
55 For example, the argument that a given act of emitting CO_2, such as a single Sunday afternoon joyride, does not on its own cause or meaningfully contribute to climate change. See Sinnott-Armstrong 2005.
56 Swann et al. 2012.
57 This in turn raises an interesting question as to whether or not greenhouse gas emissions may in fact now be considered a form of geoengineering since we have full knowledge that increasing atmospheric concentrations of these gases will change the global climate. I will set aside this question here, though it merits further consideration.
58 Cusack et al. 2014.
59 Cusack et al. 2014.
60 For a discussion of the potential for carbon sequestration to play a role in a mitigation strategy of bringing atmospheric CO_2 down to 350 ppm, including the risks inherent to the relevant strategies, see Wennersten et al. 2014.
61 Shepherd et al. 2009.
62 Note that worries about parties unilaterally implementing geoengineering technologies cannot be addressed by the Catastrophic Precautionary Decision-Making Framework, since such a possibility would already be in violation of the Catastrophic Precautionary Principle.
63 For another perspective on why the possibility of "climate emergencies" do not yet justify geoengineering, see Sillmann et al. 2015.

64 Elliott 2010.
65 McKinnon 2012.
66 For a discussion of "arm the future" arguments for sulfate injection, which may apply to geoengineering techniques more broadly, see Gardiner 2010.
67 McKinnon 2012, 137.

References

Adger, W.Neil, IreneLorenzoni, and Karen L. O'Brien, eds. 2010. *Adapting to Climate Change: Thresholds, Values, Governance.* Cambridge: Cambridge University Press.

Anderson, Soren, and Richard Newell. 2004. "Prospects for Carbon Capture and Storage Technologies." *Annual Review of Environment and Resources* 29(1): 109–142. doi:10.1146/annurev.energy.29.082703.145619.

Barnett, Jon. 2010. "Adapting to Climate Change: Three Key Challenges for Research and Policy – An Editorial Essay." *Wiley Interdisciplinary Reviews: Climate Change* 1(3): 314–317. doi:10.1002/wcc.28.

Barros, V.R., C.B. Field, D.J. Dokken, M.D. Mastrandrea, K.J. Mach, T.E. Bilir, M. Chatterjee, et al., eds. 2014. *IPCC, 2014: Climate Change 2014: Impacts, Adaptation, and Vulnerability. Part B: Regional Aspects. Contribution of Working Group II to the Fifth Assessment Report of the Intergovernmental Panel on Climate Change.* Cambridge: Cambridge University Press.

Bodansky, Daniel. 1996. "May We Engineer the Climate?" *Climatic Change* 33(3): 309–321. doi:10.1007/BF00142579.

Cancun Adaptation Framework. 2010. http://unfccc.int/meetings/cancun_nov_2010/items/6005.php.

Crutzen, Paul J. 2006. "Albedo Enhancement by Stratospheric Sulfur Injections: A Contribution to Resolve a Policy Dilemma?" *Climatic Change* 77(3–4): 211–220.

Cusack, Daniela F., Jonn Axsen, Rachael Shwom, Lauren Hartzell-Nichols, Sam White, and Katherine R.M. Mackey. 2014. "An Interdisciplinary Assessment of Climate Engineering Strategies." *Frontiers in Ecology and the Environment* 12(5): 280–287. doi:10.1890/130030.

Dovers, Stephen R., and Adnan A. Hezri. 2010. "Institutions and Policy Processes: The Means to the Ends of Adaptation." *Wiley Interdisciplinary Reviews: Climate Change* 1(2). doi:10.1002/wcc.29.

Dow, Kirstin, Roger E. Kasperson, and Maria Bohn. 2006. "Exploring the Social Justice Implications of Adaptation and Vulnerability." In *Fairness in Adaptation to Climate Change*, ed. W.N. Adger, Jouni Paavola, Saleemul Huq, and M.J. Mace, 79–96. Cambridge: MIT Press.

Eakin, Hallie, Emma L. Tompkins, Donald R. Nelson, and John M. Anderies. 2010. "Hidden Costs and Disparate Uncertainties: Trade-Offs in Approaches to Climate Policy." In *Adapting to Climate Change: Thresholds, Values, Governance*, ed. W.N. Adger, Irene Lorenzoni, and Karen L. O'Brien, 212–226. Cambridge: Cambridge University Press.

Elliott, Kevin. 2010. "Geoengineering and the Precautionary Principle." *International Journal of Applied Philosophy* 24(2): 237–253.

Fankhauser, Samuel. 2010. "The Costs of Adaptation." *Wiley Interdisciplinary Reviews: Climate Change* 1(1): 23–30. doi:10.1002/wcc.014.

Field, C.B., V.R. Barros, D.J. Dokken, K.J. Mach, M.D. Mastrandrea, T.E. Bilir, M. Chatterjee, et al., eds. 2014. *IPCC, 2014: Climate Change 2014: Impacts, Adaptation, and Vulnerability. Part A: Global and Sectoral Aspects. Contribution of Working Group II to the Fifth Assessment Report of the Intergovernmental Panel on Climate Change*. Cambridge: Cambridge University Press.

Fleming, James Rodger. 2010. *Fixing the Sky: The Checkered History of Weather and Climate Control*. New York: Columbia University Press.

Gardiner, Stephen M. 2010. *Is "Arming the Future" with Geoengineering Really the Lesser Evil? Some Doubts about the Ethics of Intentionally Manipulating the Climate System*, ed. Stephen M. Gardiner, Simon Caney, Dale Jamieson, and Henry Shue. Oxford: Oxford University Press.

Gardiner, Stephen M. 2011. *A Perfect Moral Storm: The Ethical Tragedy of Climate Change*. Oxford: Oxford University Press.

Hamilton, Clive. 2013. *Earth Masters: The Dawn of the Age of Climate Engineering*. New Haven: Yale University Press.

Hansen, James, Makiko Sato, Pushker Kharecha, David Beerling, Robert Berner, Valerie Masson-Delmotte, Mark Pagani, Maureen Raymo, Dana L. Royer, and James C. Zachos. 2008. "Target Atmospheric CO_2: Where Should Humanity Aim?" *The Open Atmospheric Science Journal* 2(1). http://benthamopen.com/ABSTRACT/TOASCJ-2-217.

Hansen, James, Pushker Kharecha, Makiko Sato, Valerie Masson-Delmotte, Frank Ackerman, David J. Beerling, Paul J. Hearty, et al. 2013. "Assessing 'Dangerous Climate Change': Required Reduction of Carbon Emissions to Protect Young People, Future Generations and Nature." *PloS One* 8(12): e81648. doi:10.1371/journal.pone.0081648.

Hansen, J., M. Sato, P. Hearty, R. Ruedy, M. Kelley, V. Masson-Delmotte, G. Russell, et al. 2015. "Ice Melt, Sea Level Rise and Superstorms: Evidence from Paleoclimate Data, Climate Modeling, and Modern Observations that 2°C Global Warming is Highly Dangerous." *Atmospheric Chemistry and Physics Discussions* 15(14): 20059–20179. doi:10.5194/acpd-15-20059-2015.

Hartzell-Nichols, Lauren. 2011. "Responsibility for Meeting the Costs of Adaptation." *Wiley Interdisciplinary Reviews: Climate Change* 2(5): 687–700. doi:10.1002/wcc.132.

House, Kurt Zenz, Antonio C. Baclig, Manya Ranjan, Ernst A. van Nierop, Jennifer Wilcox, and Howard J. Herzog. 2011. "Economic and Energetic Analysis of Capturing CO_2 from Ambient Air." *Proceedings of the National Academy of Sciences of the United States of America* 108(51): 20428–20433. doi:10.1073/pnas.1012253108.

Jamieson, Dale. 2005. "Adaptation, Mitigation, and Justice." In *Perspectives on Climate Change: Science, Economics, Politics, Ethics*, ed. Walter Sinnott-Armstrong and Richard B. Howarth, 217–248. New York: Elsevier.

Jamieson, Dale. 2013. "Some Whats, Whys and Worries of Geoengineering." *Climatic Change* 121(3): 527–537. doi:10.1007/s10584-013-0862-9.

Jamieson, Dale. 2014. *Reason in a Dark Time: Why the Struggle Against Climate Change Failed – and What it Means for Our Future*. Oxford: Oxford University Press.

Keith, David W. 2000. "Geoengineering the Climate: History and Prospect." *Annual Review of Energy and the Environment* 25(1): 245–284.

Landry, M.R., M.E. Ondrusek, S.J. Tanner, S.L. Brown, J. Constantinou, R.R. Bidigare, K.H. Coale, and S. Fitzwater. 2000. "Biological Response to Iron

Fertilization in the Eastern Equatorial Pacific (IronEx II). I. Microplankton Community Abundances and Biomass." *Marine Ecology Progress Series* 201: 27–42. doi:10.3354/meps201027.

Lenton, Timothy M. 2011. "Beyond 2°C: Redefining Dangerous Climate Change for Physical Systems." *Wiley Interdisciplinary Reviews: Climate Change* 2(3): 451–461. http://doi.wiley.com/10.1002/wcc.107.

McKinnon, Catriona. 2012. *Climate Change and Future Justice: Precaution, Compensation, and Triage.* New York: Routledge.

Mitchell, Daniel, Rachel James, Piers M. Forster, Richard A. Betts, Hideo Shiogama, and Myles Allen. 2016. "Realizing the Impacts of a 1.5°C Warmer World." *Nature Climate Change.* doi:10.1038/NCLIMATE3055.

National Academy of Sciences. 1992. *Policy Implications of Greenhouse Warming: Mitigation, Adaptation, and the Science Base.* Washington, DC.

Nelson, Donald R. 2010. "Conclusions: Transforming the World." In *Adapting to Climate Change: Thresholds, Values, Governance*, ed. W. Neil Adger, Irene Lorenzoni, and Karen L. O'Brien, 491–500. Cambridge: Cambridge University Press.

NOAA (National Oceanic and Atmospheric Administration). 2015. "Global Greenhouse Gas Reference Network." Accessed August 11. www.esrl.noaa.gov/gmd/ccgg/trends/global.html.

Nordhaus, William. 2013. *The Climate Casino: Risk, Uncertainty, and Economics for a Warming World.* New Haven: Yale University Press.

Orlove, Ben. 2001. *The Past, the Present and Some Possible Futures of Adaptation.* Cambridge: Cambridge University Press.

Paavola, Jouni, and W. Neil Adger. 2006. "Fair Adaptation to Climate Change." *Ecological Economics* 56(4): 594–609. doi:10.1016/j.ecolecon.2005.03.015.

Pacala, S., and R. Socolow. 2004. "Stabilization Wedges: Solving the Climate Problem for the next 50 Years with Current Technologies." *Science (New York, N.Y.)* 305(5686): 968–972. doi:10.1126/science.1100103.

Pelling, Mark. 2011. *Adaptation to Climate Change: From Resilience to Transformation.* Routledge.

Peterson, Garry. 2010. "Ecological Limits of Adaptation to Climate Change." In *Adapting to Climate Change: Thresholds, Values, Governance*, ed. W. Neil Adger, Irene Lorenzoni, and Karen L. O'Brien, 25–41. Cambridge: Cambridge University Press.

Rockström, Johan, Will Steffen, Kevin Noone, Asa Persson, F. Stuart Chapin, Eric F. Lambin, Timothy M. Lenton, et al. 2009. "A Safe Operating Space for Humanity." *Nature* 461(7263): 472–475. doi:10.1038/461472a.

Schneider, Stephen H. 1996. "Geoengineering: Could? Or Should? We Do It?" *Climatic Change* 33(3): 291–302. doi:10.1007/BF00142577.

Shepherd, John, Ken Caldeira, Peter Cox, Joanna Haigh, David Keith, Brian Launder, Georgina Mace, et al. 2009. "Geoengineering the Climate: Science, Governance and Uncertainty." The Royal Society. London.

Sillmann, Jana, Timothy M. Lenton, Anders Levermann, Konrad Ott, Mike Hulme, François Benduhn, and Joshua B. Horton. 2015. "Climate Emergencies Do Not Justify Engineering the Climate." *Nature Climate Change* 5(4): 290–292. doi:10.1038/nclimate2539.

Sinnott-Armstrong, Walter. 2005. "It's Not My Fault: Global Warming and Individual Moral Obligations." In *Perspectives on Climate Change: Science,*

Economics, Politics, Ethics, ed. Walter Sinnott-Armstrong and Richard B. Howarth, 221–253. Amsterdam: Elsevier.

Stern, Nicholas. 2015. *Why are We Waiting? The Logic, Urgency, and Promise of Tackling Climate Change.* Cambridge: MIT Press.

Stocker, T.F., D. Qin, G.-K. Plattner, M. Tignor, S.K. Allen, J. Boschung, A. Nauels, Y. Xia, V. Bex, and P.M. Midgley, eds. 2013. *IPCC, 2013: Climate Change 2013: The Physical Science Basis. Contribution of Working Group I to the Fifth Assessment Report of the Intergovernmental Panel on Climate Change.* Cambridge: Cambridge University Press. doi:10.1017/CBO9781107415324. Summary.

Swann, Abigail L.S., Inez Y. Fung, and John C.H. Chiang. 2012. "Mid-Latitude Afforestation Shifts General Circulation and Tropical Precipitation." *Proceedings of the National Academy of Sciences of the United States of America* 109(3): 712–716. doi:10.1073/pnas.1116706108.

UNFCCC. 1992. *United Nations Framework Convention on Climate Change.* http://unfccc.int/resource/docs/convkp/conveng.pdf.

UNISDR (United Nations Office for Disaster Risk Reduction). 2005. *Hyogo Framework for Action 2005–2015: Building the Resilience of Nations and Communities to Disasters.* www.unisdr.org/files/1037_hyogoframeworkforactionenglish.pdf.

United Nations. 2015. "Adoption of the Paris Agreement." Conference of the Parties on Its Twenty-First Session21932 (December): 32.

Vaughan, Naomi E., and Timothy M. Lenton. 2011. "A Review of Climate Geoengineering Proposals." *Climatic Change* 109(3–4): 745–790.

Vuuren, Detlef P., Jae Edmonds, Mikiko Kainuma, Keywan Riahi, Allison Thomson, Kathy Hibbard, George C. Hurtt, et al. 2011. "The Representative Concentration Pathways: An Overview." *Climatic Change* 109(1–2): 5–31. www.springerlink.com/index/10.1007/s10584-011-0148-z.

Wagner, Gernot, and Martin L. Weitzman. 2015. *Climate Shock: The Economic Consequences of a Hotter Planet.* Princeton: Princeton University Press.

Wayne, G.P. 2013. "The Beginner's Guide to Representative Concentration Pathways." *Skeptical Science.* www.skepticalscience.com/rcp.php.

Wennersten, Ronald, Qie Sun, and Hailong Li. 2014. "The Future Potential for Carbon Capture and Storage in Climate Change Mitigation – an Overview from Perspectives of Technology, Economy and Risk." *Journal of Cleaner Production* 103 (September): 724–736. doi:10.1016/j.jclepro.2014.09.023.

Appendix
Getting Around the Non-Identity Problem

This Appendix expands on the discussion in Chapter 1 about the philosophical challenges to understanding harmfulness across generations. The first section offers a more technical discussion of the non-identity problem, which is only alluded to in Chapter 1. The second section then picks up where Chapter 1 left off, offering a more technical argument for avoiding the notion that climate change harms individuals. The final section considers (though rejects) an alternative view of harm to highlight the distinct features of my view and what is at stake.

1 The Challenge of the Non-identity Problem

A philosophical puzzle, known at the non-identity problem, further complicates how we can understand the ways in which climate change is and will be harmful because of its intergenerational nature. This puzzle identifies an implication of actions that harmfully affect future people, of which climate-affecting activities are a prime example. It turns out that the very same actions that affect climate will also affect the identity of who lives in the future and hence who will be harmfully affected by climate change. This makes it challenging to understand whether or how future people will be harmed by climate change since they will owe their existence to the very actions that led to their being harmfully affected.

Derek Parfit explains the non-identity problem by illustrating how we can affect the personal identity of future people.[1] He asks two questions to distinguish three different kinds of choices we make, each of which has conceptual and ethical implications. These are "[w]ould all and only the same people ever live in both outcomes?" and "[w]ould all and only the same number of people ever live in both outcomes?"[2] Key to understanding the different kinds of choices we make is the assumption that the cells out of which a person develops determine who a person is. This view appeals to the *Time-Dependence Claim* that a person would never have existed if she had not been conceived when she was in fact conceived.[3]

When the same people would live with either outcome of a choice, the decision is what Parfit calls a *same people choice*. In general, these are choices that do not significantly affect the future. A clear example of a same people choice is your decision to brush your teeth or not this morning. Whether you brushed your teeth or not mostly likely did not affect who will live in the future. When the same number but different people live with either outcome, the decision is a *same number choice*. An example of a same number choice is a woman's successful choice to use birth control to determine when she will have a set number of children. By using birth control a woman affects the personal identity of her offspring, at least in part, by determining when her children are conceived and hence their genetic makeup. Finally, when different people and a different number of people live with either outcome, the decision is a *different number choice*. These are choices that very much affect the future. A straightforward example of a different number choice is a woman's decision to have regular sex without using any form of birth control. Her choice affects how many children she will have, though she cannot predict the outcome of this choice in advance.[4]

Parfit argues that different kinds of choices require different moral considerations. He points out that we often apply our intuitions about same people choices to same number or even different number choices. This is not appropriate, however, because of the issue of personal identity. The fact that our personal identity is in part tied to our genetics has significant implications. Parfit poses a powerful question when he says, "how many of us could truly claim, 'Even if railways had never been invented, I would still have been born'?"[5] On reflection, the answer to his question is virtually "none." The invention of the railroad was significant enough to have affected the lives of many of the people who existed at the time and consequently who their offspring were, leading up to our current population. Had the railroad not been invented and introduced to society when it was, history would not have led to the existence of precisely the same collection and number of people.

Climate-affecting activities and policies have the potential to be as or even more significant in the course of history as the invention of the railroad. Policies geared towards the development of new technologies, for example, could lead to advances or changes as or more significant than this development. If this happens, future people 200 years from now will be able to answer the question, "even if such and such technology had never been invented, would you have still been born?" in the negative, the way we today answer the question about the railroad. Similarly, if policies allowing the status quo to continue are adopted, future people 200 from now will owe their existence to society's historical reliance on fossil fuels. Looking back in time we can also see that policies supporting the Industrial Revolution and the development of a fossil fuel-based economy led to the existence of precisely those who exist today. Not only would we not have existed if the railroad had not been invented, but we most certainly would not exist had the Industrial Revolution not occurred. Both of these events affected who came into existence.

By identifying the relationship between personal identity and choices that affect the future, Parfit exposes that fact that many decisions affect who will exist. When a decision affects who will exist, however, it is unlikely that different actions will result in the same number of people. Even if a decision causes the existence of just one extra person or causes there to be one less person it is a different number choice. Different number choices do not require a total change in the number and personal identity of future persons. When the quality of people's lives is changed, the timing and parenthood of conceptions is changed. Therefore most of our choices that affect the lives of even small groups of individuals are different number choices. This means that most societal-level choices are different number choices.

Casper Hare sums up this point nicely when he says, "[g]iven that world history is a large and encompassing thing, it seems likely that most decisions that affect who exists will reverberate through it for many generations and unlikely that, when all is said and done, the numbers of people who ever exist will turn out the same whatever we decide."[6] In this way, societal-level choices are different number choices that affect the number and personal identity of future people. One of the most significant challenges this fact poses is the widely discussed non-identity problem.[7] The non-identity problem stems from the fact that our choices can affect the future in seemingly bad ways despite the fact that we cannot harm specific individuals in causing to exist. That is, if we assume that causing to exist cannot be worse for someone who has a life worth living, we cannot make a future person worse off and hence cannot harm her. Parfit states this as the problem of identifying the moral difference between outcomes that are worse for no specific individuals.[8]

Parfit argues that choices that affect the personal identity of future people require different moral considerations than those that do not. He points out that we often apply our intuitions about choices that do not affect personal identity to those that do, but this is not appropriate. It may often seem that different number choices that have obvious harmful effects are bad because they appear to harm future people, but the non-identity problem challenges the idea that such choices in fact harm future people because whoever exists in the future will owe their existence to the same actions that created the (harmful) conditions in which they live. Whoever lives will owe their existence to our choices and hence as individuals cannot have been made worse off by those choices. This means that we cannot lean on our intuitive notion about the wrongfulness of harming to explain why it is morally bad to harmfully affect the future people whose existence depends on our actions.

Key to the present discussion is that the non-identity problem challenges the way in which we colloquially talk about harming future generations. The time-lagged nature of climate change means that climatic effects are owed in part to GHGs emitted tens, hundreds, or even thousands of years prior to the effect. If to harm someone is to make her worse off in some way, then it seems we cannot harm future people whose existence we in part determine. The idea that climate change is harmful to future people is threatened by the non-identity problem since it is no longer clear that the future people can be harmed at all.

Put another way, none of us can claim that the invention of the railroad harmed us, even if railroads have somehow negatively affected us. Even someone who was disfigured or who lost her family in a train wreck cannot claim to have been harmed by the invention of the railroad because she would not exist and therefore would not have been harmfully affected had the railroad not been invented.[9] This illustrates that one of the most significant aspects of the temporal delay between actions that contribute to climate change and their effects is the fact that these very same actions affect who will live in the future. The very idea that we can harm future people is threatened by the non-identity problem since it is no longer clear that specific future individuals can be harmed at all.

However, I should point out that not everyone is so concerned about the puzzle the non-identity problem presents. Edward Page, for example, after offering an in-depth discussion of the non-identity problem and responses to it, in particular as these relate to intergenerational justice, takes the "no difference view."[10] Much like Parfit, he argues, "the problem should inspire us to think seriously about the theoretical basis for the responsibilities to which many of us are already intuitively committed."[11] While I agree with this general sentiment, in order even to make sense of how climate change is harmful, it is important that we have a way to talk about harming relationships that span generations and the non-identity problem challenges this possibility. This does not mean that we cannot *wrong* future people. Rather it means that the moral concepts we attach to here-and-now harming do not apply in cases in which our actions affect who will live in the future. Making sense of the way in which contributing to climate change is harmful to future people will provide a means of clearly discussing the nature of our relationship to future people. This in turn may help us see when and why it is wrong to harmfully affect future generations in spite of the puzzling conclusion that we cannot harm in the usual way in causing to exist.

2 When Harming is not Harming – A Way Forward

In Chapter 1 I argued that it is helpful to focus on the way in which climate change is harmful rather than trying to make our usual concept of harm conform to the intergenerational context in which it operates. Casper Hare's use of the *de re* (of the thing) vs. *de dicto* (of the word) distinction can further help clarify how we may understand the possibility of inter-generational harming that expands on this idea.[12] Hare identifies two kinds of harm: one of which is the traditional kind of harm, and one of which is essentially impersonal. He uses a helpful joke to illustrate this distinction: Zsa Zsa has found a way to keep her husband young and healthy. The joke is that the source of Zsa Zsa's proverbial fountain of youth is that she gets remarried every five years. Zsa Zsa's method for keeping her husband young and healthy promotes these good qualities in an impersonal *de dicto* way. Promoting these qualities in a *de re* way would require that she always have the same husband because *de re* goodness attaches to a particular person. While Hare's chapter focuses on *de dicto* goodness, it is *de dicto* badness or wronging that is relevant to the present account. Adapting Hare's definitions we get:

> *De re worse*: where S1 and S2 are states of affairs, S1 is *de re* worse for the health of ___ than S2, when the thing that is actually ___ is sicker in S1 than in S2.
> *De dicto worse*: where S1 and S2 are states of affairs, S1 is *de dicto* worse for the health of ___ than S2, when the thing that is ___ in S1 is sicker in S1 than the thing that is ___ in S2 is in S2.[13]

As Hare rightly notes, *de dicto* badness does not always matter morally, but it sometimes does. *De dicto* badness can apply to different number choices where "normal" *de re* badness cannot. As Hare says, "there are non-identity cases ... about which the de dicto betterness account gives clear answers, though they may involve actions that affect how many people ever exist."[14] Choices may be *de dicto* worse for future people (the specific people who come to exist in the future) even though they affect the personal identity and number of future people.

A version of the now classic example of a mother who knowingly gives birth to a child with deformities or defects illustrates this point. Imagine Katie conceives a child against the recommendation of her doctor while on a certain medication that causes her child to have a birth defect. Had Katie waited one month to conceive until she was off the medication, she would have had a normal child. Feinberg says of the state of the "defective child" if she is conceived, "I prefer to call it, therefore, a *harmful condition* rather than a harmed condition."[15] That is, he says she is not in a *harmed condition* because there was no prior act of harming that caused her to be in the state she is in. Hare points out that we are right to think that Katie "makes things de dicto worse for the health of her future child, and this is something she should have been concerned to avoid."[16] Before deciding to conceive there was no way for Katie to express *de re* concern for her child since she could not at that point know the personal identity of her child. Yet it is appropriate to be concerned about one's future children; this concern is a type of *de dicto* concern. Bringing Feinberg's and Hare's accounts together gives us a more nuanced way of talking about this kind of concern. *De dicto* harming future people involves putting them into a *harmful*, though not harmed, *condition*.

Extending Hare's account we can understand the way in which actions can *de dicto* harm future people.

> *De re harming*: X *de re* harms Y when X makes Y worse off (i.e., puts her in a harmed condition).
> *De dicto harming*: X *de dicto* harms Y when X's action puts Y in a harmful condition, though Y is not made worse off or *de re* harmed by X's action (i.e., is not in a harmed condition).

Katie *de dicto* harms her future child when she conceives while on the drug, though she does not make *him* – the particular person who will be her son – worse off. She does this by imposing a harmful condition on her future child, though she cannot cause him to be in a harmed condition. Ignoring her doctor's recommendation and trying to conceive while on the medication amounts to Katie knowingly *de dicto* harming her future child, whoever he turns out to be. From his perspective, Katie's son may be harmfully affected and made worse off at some point during his life, but he will not be able to claim that Katie made him worse off for he would not have existed had she not conceived him when she did. While we may think Katie's act of *de dicto* harming is wrong since parents have especially strong *de dicto* obligations not to harm their children, whoever they turn out to be, not all acts of *de dicto* harming are necessarily wrong. For example, there might be circumstances in which no matter when a person conceives, her children will necessarily be brought into a harmful condition. In such a case it may not be wrong to conceive if, for example, despite experiencing some harmful conditions any child so conceived will nonetheless live a life worth living.

One may think that while we can apply this concept of *de dicto* harming to a choice that affects the identity of just one person, we cannot apply it to choices that affect how many people will live in the future, but this is mistaken. Altering the example slightly opens the door to this possibility: imagine that, unbeknownst to her, the same Katie discussed above has a propensity to have fraternal twins. To simplify things, further imagine that she only intends to have one pregnancy. If this is the case, her choice of whether or not to delay conceiving is at least potentially a different number choice. Nonetheless, we can still say that she *de dicto* harms her future child(ren) if she tries to conceive while on the medication since this will be *de dicto* harmful to her future child(ren) whoever he or they turn out to be. The difference between the original and this amended case is that it is not immediately obvious whether it is wrong of Katie to *de dicto* harm her future child(ren) in the different number case. As Hare recognized, not all cases of *de dicto* harming will be judged to be wrong, but the concept of *de dicto* harming applies as well to choices that affect the number of future people as it does to choices that only affect the personal identity of a given set of future people.

All of this suggests that while future people cannot be *de re* harmed by actions that led to their being in such a state since they will owe their existence to such actions, they can still be in a harmful condition that was not the result of an act of harming. Assuming that climate change will negatively impact some individuals, we can understand acts that contribute to climate change as harmfully affecting or *de dicto* harming future people.[17]

It is important to be clear that the context in which Feinberg uses the notion of harmful conditions is much more limited than the present context. By bringing together Feinberg's and Hare's accounts I am pushing them both further than either author has taken their own ideas. Feinberg says that cases of harmful conception like Katie's are the only cases where a person is "put in a harmful condition by the very act that brings him into existence, and the only example where determinations of harm require comparison of a given condition with no existence at all."[18] Feinberg appears to assume that the only cases where the same action that causes an individual to exist also causes her to live in harmful conditions will occur on the level of individuals conceiving or not conceiving in particular instances. While it is true that there is something special about cases in which the harm relationship is relative to the fact that the person would not exist were it not for the very thing that caused them to exist in harmful conditions, our actions also causally contribute to the existence of (distant) future people in ways that cause them to exist in harmful conditions where our actions are not those of harmful conceptions.

The implication of this is that Feinberg's identification of the fact that we can create harmful conditions in which a person lives without and in spite of the fact that we cannot *de re* harm her is much more significant than he realized. It is not merely that so-called wrongful conception is harmful for the person it creates, but that our actions can be harmful to the people to whose existence we causally contribute, despite the fact that we cannot *de re* harm them. At the same time, expanding the interpretation of harmful conditions in this way makes it so that not all harmful conditions will be the result of acts of wrongdoing. This aligns with the colloquial application of harmfulness language to non-volitional harms and to the sense in which there can be *de dicto* harms resulting from choices that affect who will live in the future. A tornado can harmfully affect people, but it cannot *de re* harm them, and it certainly does not *wrong* anyone. Similarly, a hurricane that was causally affected by climate change and hence the actions of people in the past can harmfully affect individuals though it does not *de re* harm them. In this latter case we can say that past people contributed to the *de dicto* harming of those affected by the hurricane even though their climate-affecting activities (e.g., GHG emissions) may also have contributed to the victim's very existence. We can then explain how though the hurricane did not wrong anyone, those who contributed to the conditions that led to the hurricane may have done so. In such a case we will have to provide an explanation of this wrongdoing that does not appeal to standard *de re* harm.

This conceptual mapping aligns with the colloquial application of harmfulness language to harmful conditions (e.g., those caused by a falling tree branch) and to the sense in which there can be *de dicto* harms resulting from choices that affect who will live in the future. Hare's and Feinberg's distinctions can help us see that there is a distinction to be made between *de re* and *de dicto* harming relationships, only the latter of which can hold between generations. The present account, which aligns with but goes much further than Feinberg's and Hare's accounts, allows us to understand the way in which climate change is harmful without having to completely abandon the idea that we can harm future people. Yet it forces us to acknowledge that the ways in which we can harm future people are very different from the ways in which we can harm existing people. Most importantly, it will be much harder to sort out when and why *de dicto* harmful acts are wrong than it is to sort out when and why *de re* harmful acts are wrong. This conclusion reinforces the challenge of intergenerational ethics while giving us a vocabulary for talking about the way in which climate change is harmful.

3 An Alternative View

There is a vast literature on the non-identity problem and a multitude of responses and solutions to it.[19] Here I will consider just one especially illustrative view so as to reinforce the interpretation of intergenerational harming presented above. Elizabeth Harman has a different view of what harm is from that which has been presented so far.[20] In her view, we can in fact harm future people and not in a merely *de dicto* way. Harman asserts that a sufficient condition for harm is that "[a]n action harms a person if the action causes pain, early death, bodily damage, or deformity to her, even if she would not have existed if the action had not been performed."[21] Harman's view might thus be understood as a kind of threshold understanding of harm whereby, as Lukas H. Meyer and Dominic Roser say, "an action harms a person if as a consequence of that action the person falls under a normatively defined threshold."[22] To harm someone on a threshold view is to make him worse off relative to a normatively defined objective state, which for Harman is a normal healthy bodily state.[23] On Harman's view we harm future people whenever we cause them to be in a state that is below the threshold for a healthy bodily state.[24]

So for Harman, that something is a harm is a reason against it, but other reasons can outweigh the negative reason provided by the harm.[25] Essential to Harman's view is that benefits to future people offer reasons in favor of actions, but reasons against harm are "morally serious" and hard to outweigh.[26] Critical to Harman's argument is her position, "that an action may be wrong *in virtue of harming* even though it makes a person better off than she would otherwise be."[27] She articulates this more clearly when she claims that:

> reasons against harm are so morally serious that the mere presence of greater benefits to those harmed is not in itself sufficient to render the harms permissible: when there is an alternative in which parallel benefits can be provided without parallel harms, the harming action is wrong.[28]

In a more recent chapter, Harman clarifies that in non-identity cases, the benefits of existence are "*ineligible* to justify the harm if *failing to perform the action* would similarly benefit someone."[29]

The commonly used case of surgery illustrates Harman's basic view. Suppose Susie is having her tonsils removed because she suffers from frequent bouts of tonsillitis. Along with Feinberg and others, I do not believe that the surgeon harms Susie when she cuts away her tonsils if the removal of Susie's tonsils is in her interests; a tonsillectomy does not make Susie worse off. Harman, however, does believe the surgeon harms Susie when she cuts into her, since in so doing she injures Susie's body. Nonetheless, Harman would say that it is permissible for the surgeon to harm Susie in this way since she has good reasons for doing so, namely to promote Susie's overall health and well-being. So on Harman's account this harm is not morally bad or wrong.

The implication of Harman's view that is relevant to this discussion is that she argues that we can in fact harm future people, that the non-identity problem is only a problem because it confuses what constitutes harming. Harman avoids the non-identity problem by making the point of comparison for determining when something constitutes harm an ideal healthy bodily state. In so doing she avoids the problem that future people cannot be harmed because they cannot be made better or worse off, as is usually required to judge whether something is a (*de re*) harm or benefit. For Harman, *any* act causing bodily injury is an act of harming. It follows that any act that causes a future person to suffer a bodily injury would thus also be an act of harming. At the same time, Harman's claim about the ineligibility of benefits that could accrue to a different future individual in non-identity cases allows her to assess the moral status of acts of inter-generational harming. Without this additional claim her view would otherwise imply that harming future people is never wrong because the benefits of existence will always outweigh the badness of the physical harms future people will suffer. Harman's claim about ineligible reasons allows her to begin to offer an account of when harming future people is wrong despite the benefits of existence.

The key difference between my view and Harman's is what we take to be the relevant point of comparison when assessing whether an act is one of harming. For Harman the point of comparison is always a healthy bodily state. In my view, the point of comparison is the actor's view of the state of the person being affected because it does not make sense to say that I harm a future person who endures bodily injury as a result of my action if this same action is on the whole not harmful to this person. Harman can claim we can harm future people because we will always be able to compare the state of an individual to that of an abstract healthy bodily state. However, I require that an actor would have to contribute causally to an individual being made worse off in order for her act to be one of harm, which is not possible in non-identity cases. Nonetheless, I also argue that one can always be harmfully affected from his own perspective, even if some of the causal forces at work contributed to his existence. This is why I claim that climate change harmfully affects future people since some future people will be made worse off from their own perspective despite the fact that they were not made worse off by those who contributed to their being in this state (e.g., those emitting GHGs). My view of what constitutes a harmful condition is similar to Harman's account of a harmed condition, the difference being that for Harman a harmful condition is always a harmed condition if it was caused by a harmful act (as opposed to a natural event).

Harman's interpretation of harm enables her to weigh in on the question of when it is wrong to harm future people, but it does so at the expense of our usual way of thinking about harming our contemporaries. She conflates harming with injuring. I do not want to let go of the idea that to harm is to make someone worse off (though I am happy to allow for different interpretations of what kinds of worsening count). Usually injuring someone does harm them, but not always. To say that we harm our children when we give them a shot of antibiotics with the intention of curing their staph infection would be to undermine the most basic meaning of harm. Of course a shot may temporarily injure a child, but it does not harm her. Maybe we can say that the shot harmfully affects the child since from her perspective it seems like her network of interests is adversely affected, she feels like she is made worse off,[30] but from our perspective, say as her parents and doctor, we know that the shot does not in fact make her worse off. To say that we harmed her takes away the normative content of what it is to harm. Harman accepts this; I think we should not.[31] Maintaining the standard account of (*de re*) harm and understanding intergenerational harming in the way I have suggested will enable us to have a common ground from which to explore the challenging questions about when and why it is wrong to harmfully affect future generations. We will have to accept that our reasons in non-identity cases will be different from those in cases that do not affect the identity of future persons.

4 Conclusion

So again, how is climate change harmful? People will be harmfully affected by climate change, though no one will have directly harmed them in the standard *de re* way. The view presented here enables us to talk about the way in which climate change is and will likely be harmful without abandoning our concept of what it means to harm. Accepting this view will require that we are more careful when we talk about the ways in which we harm future generations; we can only *de dicto* harm or harmfully affect future people. While we certainly do not have to go around distinguishing *de re* from *de dicto* harm in our everyday lives or even every time we talk about harmful climate impacts, it is important that we understand why the ethics of climate change is so complex. While we can quite simply say that climate change is bad because it will be harmful, we cannot appeal to an overly simplistic account of *de re* harm to explain what this entails about the wrongness of our climate-affecting activities or what we should do to address climate change. Our explanation of why we should do something about climate change, why we should try to avoid its harmful impacts, must be responsive to the complex ways in which climate change is harmful. This is precisely the challenge this book as a whole aims to meet.

Notes

1 Parfit 1984.
2 Parfit 1984, 295.
3 Parfit 1984, 351.
4 It is worthwhile to note that even the same people and same number choice examples given here could turn out to be different number choices if, for example, whether you brushed your teeth affected whether or not you had sex with your partner or if we consider the fact that children conceived at different times may themselves be more or less likely to have more or fewer children.
5 Parfit 1984, 290.
6 Hare 2007, 520.
7 See Boonin 2014.
8 Parfit 1984, 378.
9 Obviously many other causal processes would have been involved here, but the invention of the railroad was certainly a precondition for her having been affected in this way. Similarly, as discussed in Chapter 1, climatic effects will almost never be the sole cause of harmful outcomes, though they will play a causal role in many harmful outcomes.
10 (Page 2006)
11 (Page 2006, 165.
12 Hare 2007.
13 Adapted from Hare 2007, 514.
14 Hare 2007, 521.
15 Feinberg 1990, 27.
16 Hare 2007, 516.
17 Note that I am not claiming anything about responsibility here. It may turn out that nations, rather than individuals, should be held responsible for the *de dicto* harmful actions of their citizens. Sinnott-Armstrong, for example, argues against individual responsibility (Sinnott-Armstrong 2005). The present account, however, does not immediately weigh in on the debate over who – nations, individuals, and/or other actors – should be understood as bearing responsibility for the *de dicto* harms caused by GHG emissions.

18 Feinberg 1990, 327.
19 One recent collection that addresses this problem and points to the relevant literature is Roberts and Wasserman 2009. For non-consequentialist responses to the non-identity problem see also Reiman 2007; Kumar 2003; Grosseries and Meyer 2012.
20 Harman 2009, 2003, 2004.
21 Harman 2004, 107.
22 Meyer and Roser 2009, 228. In their chapter, Meyer and Roser offer their own sufficientarian threshold notion of harm (Meyer and Roser 2009).
23 It is worth noting that the concept of a normal healthy bodily state will be very hard to define and defend, though I do not pursue this objection to Harman's view here.
24 On Harman's view there is an additional threshold, which she does not identify, below which harms are never justified (Harman 2009).
25 Harman 2003, 116.
26 Harman 2004, 108.
27 Harman 2003, 105.
28 Harman 2004, 93.
29 Harman 2009, 139.
30 Some may resist saying that she is in a harmful condition, since her whole network of interests is not in fact adversely affected. Whether we view the child as being harmfully affected will depend on a more precise account of what constitutes a harmful condition than I have given here.
31 Note that while I argue we should not accept Harman's account of harm, this does not mean that we must reject her entire view. Her views about when acts that, as I would say, harmfully affect future people are wrong may be able to be reinterpreted using the account of harm that I offer here. That is, some of her conclusions about non-identity cases may stand (Harman 2009).

References

Boonin, David. 2014. *The Non-Identity Problem and the Ethics of Future People.* Oxford: Oxford University Press.

Feinberg, Joel. 1990. *Harmless Wrongdoing: The Moral Limits of the Criminal Law.* New York: Oxford University Press.

Grosseries, Axel, and Lukas H. Meyer. 2012. *Intergenerational Justice*, ed. Axel Grosseries and Lukas H. Meyer. Oxford: Oxford University Press.

Hare, Casper. 2007. "Voices from Another World: Must We Respect the Interests of People Who Do Not, and Will Never, Exist?" *Ethics* 117: 498–523.

Harman, Elizabeth. 2003. "Moral Status." Massachusetts Institute of Technology.

Harman, Elizabeth. 2004. "Can We Harm and Benefit in Creating?" *Philosophical Perspectives* 18: 89–113.

Harman, Elizabeth. 2009. "Harming as Causing Harm." In *Harming Future Persons: Ethics, Genetics and the Nonidentity Problem*, ed. Melinda A. Roberts and David T. Wasserman, 137–154. New York: Springer.

Kumar, Rahul. 2003. "Who Can Be Wronged?" *Philosophy & Public Affairs* 31(2): 99–118. doi:10.1111/j.1088-4963.2003.00099.x.

Meyer, Lukas H., and Dominic Roser. 2009. "Enough for the Future." In *Intergenerational Justice*, ed. Axel Grosseries and Lukas H. Meyer, 219–248. Oxford: Oxford University Press.

Page, Edward. 2006. *Climate Change, Justice and Future Generations.* Northampton, UK: Edward Elgar Publishing.

Parfit, Derek. 1984. *Reasons and Persons.* Oxford: Clarendon Press.

Reiman, Jefferey. 2007. "Being Fair to Future People: The Non-Identity Problem in the Original Position." *Philosophy & Public Affairs* 35(1): 69–92. doi:10.1111/j.1088-4963.2007.00099.x.

Roberts, Melinda A., and David T. Wasserman, eds. 2009. *Harming Future Persons: Ethics, Genetics and the Nonidentity Problem.* Springer Science & Business Media.

Sinnott-Armstrong, Walter. 2005. "It's Not My Fault: Global Warming and Individual Moral Obligations." In *Perspectives on Climate Change: Science, Economics, Politics, Ethics,* ed. Walter Sinnott-Armstrong and Richard B. Howarth, 221–253. Amsterdam: Elsevier.

Index